IGCSE
Geography

IGCSE
Geography

**Paul Guinness
and Garrett Nagle**

HODDER
EDUCATION
AN HACHETTE UK COMPANY

Hachette UK's policy is to use papers that are natural, renewable and recyclable products and made from wood grown in sustainable forests. The logging and manufacturing processes are expected to conform to the environmental regulations of the country of origin.

Orders: please contact Bookpoint Ltd, 130 Milton Park, Abingdon, Oxon OX14 4SB. Telephone: (44) 01235 827720. Fax: (44) 01235 400454. Lines are open 9.00–5.00, Monday to Saturday, with a 24-hour message answering service. Visit our website at www.hoddereducation.co.uk

© Paul Guinness and Garrett Nagle
First published in 2009 by
Hodder Education, an Hachette UK Company,
338 Euston Road
London NW1 3BH

Impression number 6
Year   2012

Cover photo by Garrett Nagle (Blea Tarn and the Langdale Pikes, English Lake District)
Illustrations by Art Construction and Barking Dog Art
Typeset in 11/13pt Galliard BT by Pantek Arts Ltd, Maidstone, Kent
Printed in Dubai.

Dedication
To Angela, Rosie, Patrick and Bethany

A catalogue record for this title is available from the British Library

ISBN 978 0 340 97501 5

Contents

Contents

Introduction

This book has been written to help you while you study for your geography IGCSE. The examples and case studies in the book are from around the world. Geography is about people and places and we hope that you will use your home area as much as possible to add to the material in this book. We would encourage you also to keep up-to-date with geographical events – you can do this by listening to the news or reading about events in newspapers or on the Internet. Geography is happening every day, everywhere and examiners love to read about new developments – so think about your own geographical location and new geographical events.

This book has been written to follow closely the IGCSE specification. It includes a number of activities to help you succeed with the written assessment and guidance for your coursework. Below are details of the exams and assessment that you will experience. Be prepared – knowing what to expect will help you succeed in your exams. Make sure you also make the most of your teachers' experience – they are an excellent resource waiting to be tapped. Good luck and enjoy your geographical studies.

▮ Assessment

Scheme of assessment

All candidates will take Paper 1, Paper 2 and either Paper 3 or Paper 4.

Paper 1 will be answered on a separate answer paper/answer booklet. Papers 2 and 4 will consist of combined question papers and answer booklets where candidates answer in the spaces provided.

Paper 1 (1 hour 45 minutes) Candidates will be required to answer three questions (3 × 25 marks). Six questions will be set: two on each of the three themes. Questions will be structured with gradients of difficulty, will be resource-based and involve problem solving and free response writing. This paper will mainly be concerned with Assessment Objectives A, B and C: Knowledge with understanding, Analysis, and Judgement and decision making. 45% of total marks.

Paper 2 (1 hour 30 minutes) (60 marks) This paper will be taken by all candidates. Candidates must answer all the questions. The paper is based on testing the interpretation and analysis of geographical information and on the application of graphical and other techniques as appropriate. The questions will not require specific information of a place.

One question is based on a 1:25,000 or 1:50,000 topographical map of a tropical area such as Zimbabwe, the Caribbean or Mauritius. 27.5% of total marks.

Either

Paper 3 Coursework (centre-based assessment)* (60 marks) One school-based assignment will be set by teachers of up to 2000 words. 27.5% of total marks.

*Teachers may not undertake school-based assessment without the written approval of CIE. This will only be given to teachers who satisfy CIE requirements concerning moderation and they will have to undergo special training in assessment before entering candidates.

Or

Paper 4 Alternative to coursework (1 hour 30 minutes) (60 marks) Candidates answer all the questions, completing a series of written tasks based on the three themes. The questions involve an appreciation of a range of techniques used in fieldwork studies.

Questions test the methodology of questionnaires, observation, counts, measurement techniques and may involve developing hypotheses appropriate to specific topics. The processing, presentation and analysis of data will be tested. 27.5% of total marks.

▮ IGCSE Geography Revision CD-ROM

The accompanying Revision CD-ROM provides invaluable exam preparation and practice. It has two sections:

- *Revision Support* includes sample examination papers and answers.
- *Images* includes selected artwork taken from the book to help you with your studies and projects.

  Images included from Theme 1, Unit 1 are Figures 5, 10, 14, 21, 22, 30, 32, 37, 39 and 42. Images included from Theme 1, Unit 2 are Figures 5, 7, 8, 11, 16, 19 and 22. Images included from

Theme 2, Unit 1 are Figures 2, 6 and 14. Images from Theme 2, Unit 2 include Figures 4, 5, 10, 12, 14, 16, 19, 21 and 23. Images from Theme 2, Unit 3 include Figures 7, 13, 14, 15 and an extension of the table in Figure 4. Figure 6 is included from Theme 2, Unit 4. Images included from Theme 3, Unit 1 are Figures 1, 10, 11, 14, 16 and 21. Images included from Theme 3, Unit 2 are Figures 7, 10 and 18. Images included from Theme 3, Unit 3 are Figures 3 and 9. Figure 7 is included from Theme 3, Unit 4. Images included from Theme 3, Unit 5 are Figures 3 and 11.

Paul Guinness
Garrett Nagle

Population and Settlement

Population dynamics

◼ Changes in world population growth

During most of the early period in which humankind first evolved, global population was very low, reaching perhaps some 125,000 people a million years ago. Ten thousand years ago, when people first began to domesticate animals and cultivate crops, world population was no more than five million. Known as the Neolithic Revolution, this period of economic change significantly altered the relationship between people and their environments. But even then the average annual growth rate was less than 0.1 per cent per year.

However, as a result of technological advance the **carrying capacity** of the land improved and population increased. The carrying capacity is the largest population that the resources of a given environment can support. By 3500BC global population reached 30 million and by 2000 years ago, this had risen to about 250 million (Figure 1).

Demographers (people who study human populations) estimate that world population reached 500 million by about 1650. From this time population grew at an increasing rate. By 1800 global population had doubled to reach one billion. Figure 2 shows the time taken for each subsequent billion to be reached, with the global total reaching 6 billion in 1999. It had taken only 12 years for world population to increase from 5 to 6 billion, the same timespan required for the previous billion to be added. It is estimated that the time taken for future billions to be reached will increase, with a 14-year gap until 7 billion is reached in 2013.

Figure 3 (page 2) shows population change in 2005, with a global population increase of 81 million in that year. The vast majority of this increase is in the LEDCs.

Recent demographic change

Figure 4 (page 2) shows that both total population and the rate of population growth are much higher in the less developed world than in the more developed world. However, only since the Second World War has population growth in the poor countries overtaken that in the rich. The rich countries had their period of high population

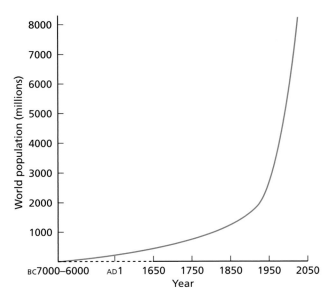

Figure 1 World population growth

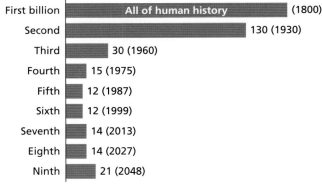

Figure 2 World population growth by each billion

Natural increase per	World	More developed countries	Less developed countries	Less developed countries (excl. China)
Year	80,794,218	1,234,907	79,559,311	71,906,587
Day	221,354	3,383	217,971	197,004
Minute	154	2	151	137

Figure 3 World population clock for 2005

growth in the nineteenth and early twentieth centuries, while for the less developed countries high population growth has occurred since 1950.

The highest ever global population growth rate was reached in the early to mid-1960s when population growth in the less developed world peaked at 2.4 per cent a year. At this time the term **population explosion** was widely used to describe this rapid growth. But by the late 1990s the rate of population growth was down to 1.8 per cent. However, even though the rate of growth has been falling for three decades (Figure 5), **demographic momentum** meant that the numbers being added each year did not peak until the late 1980s.

The demographic transformation, which took a century to complete in the developed world, has occurred in a generation in some less developed countries. Fertility has dropped further and faster than most demographers foresaw 20 or 30 years ago. Except in Africa and the Middle East, where in over 30 countries families of at least five children are the norm and population growth is still over 2.5 per cent per year, birth rates are now declining in virtually every country.

Figure 6 shows the ten largest countries in the world in terms of population size, and their population projections for 2050.

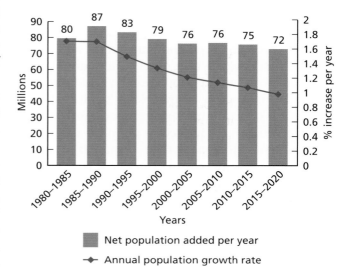

Figure 5 Population increase and growth rate, 1980–2020

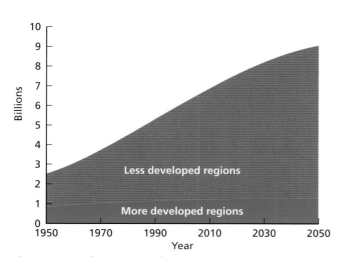

Figure 4 Population growth in more and less developed countries, 1950–2050

2007		2050	
Country	Population (millions)	Country	Population (millions)
China	1318	India	1747
India	1132	China	1437
USA	302	USA	420
Indonesia	232	Indonesia	297
Brazil	189	Pakistan	295
Pakistan	169	Nigeria	282
Bangladesh	149	Brazil	260
Nigeria	144	Bangladesh	231
Russia	142	Dem. Rep. of Congo	187
Japan	128	Philippines	150

Figure 6 The world's ten largest countries in terms of population

Activities

1 With the help of Figures 1 and 2, briefly describe the growth of human population over time.

2 Look at Figure 4. Describe the differences in population growth and projected growth in MEDCs and LEDCs between 1950 and 2050.

3 Describe the trends in a) population increase and b) population growth rate shown in Figure 5.

Factors influencing population growth

The **birth rate** is defined as the number of live births per 1000 population in a year. If the birth rate of a country is 20/1000, this means that on average for every 1000 people in this country 20 births will occur in a year. The **death rate** is the number of deaths per thousand population in a year, If the death rate for the same country is 8/1000, it means that on average for every 1000 people 8 deaths will occur. The difference between the birth rate and the death rate is the **rate of natural change**. In this case it will be 12/1000 (20/1000 – 8/1000). Figure 7 shows how much birth and death rates vary by world region.

Population change in a country is affected by a) the difference between births and deaths (natural change) and b) the balance between immigration and emigration (net migration). On Figure 8 the wavy dividing line indicates that the relative contributions of natural change and net migration can vary over time.

The **immigration rate** is the number of immigrants per 1000 population in the receiving

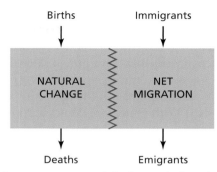

Figure 8 Input–output model of population change

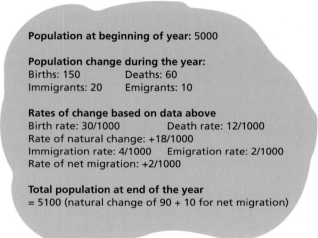

Population at beginning of year: 5000

Population change during the year:
Births: 150   Deaths: 60
Immigrants: 20   Emigrants: 10

Rates of change based on data above
Birth rate: 30/1000   Death rate: 12/1000
Rate of natural change: +18/1000
Immigration rate: 4/1000   Emigration rate: 2/1000
Rate of net migration: +2/1000

Total population at end of the year
= 5100 (natural change of 90 + 10 for net migration)

Figure 9 Pacifica: population changes in a year

country in a year. The **emigration rate** is the number of emigrants per 1000 population in the country of origin in a year. The **rate of net migration** is the difference between the rates of immigration and emigration. Figure 9 shows some calculations for the hypothetical island of Pacifica.

Factors affecting fertility

The factors affecting population change can be grouped into four categories:

● **Demographic**: Other population factors, particularly mortality rates, influence fertility. Where infant mortality is high, it is usual for many children to die before reaching adult life. In such societies, parents often have many children to compensate for these expected deaths. The **infant mortality rate** is the number of deaths of children under one year of age per 1000 live births per year.

Region	Birth rate	Death rate
World	21	9
More developed world	11	10
Less developed world	23	8
Africa	38	14
Asia	19	7
Latin America/Caribbean	21	6
North America	14	8
Oceania	18	7
Europe	10	11

Figure 7 Birth and death rates, 2007

- **Social/Cultural**: In some societies, particularly in Africa, tradition demands high rates of reproduction. Here the opinion of women in the reproductive years may have little influence weighed against intense cultural expectations. Education, especially female literacy, is the key to lower fertility (Figure 10). With education comes a knowledge of birth control, greater social awareness, more opportunity for employment and a wider choice of action generally. In some countries religion is an important factor. For example, the Muslim and Roman Catholic religions oppose artificial birth control. Most countries that have population policies have been trying to reduce their fertility by investing in birth control programmes.
- **Economic**: In many of the least developed countries children are seen as an economic asset because of the work they do and also the support they are expected to give their parents in old age. In the more developed world the general perception is reversed and the cost of the child dependency years is a major factor in the decision to begin or extend a family. Economic growth allows greater spending on health, housing, nutrition and education which is important in lowering mortality and in turn reducing fertility.
- **Political**: There are many examples in the past century of governments attempting to change the rate of population growth for economic and strategic reasons. During the late 1930s Germany, Italy and Japan all offered inducements and concessions to those with large families. In more recent years Malaysia has adopted a similar policy. However, today most governments that are interventionist in terms of fertility want to reduce population growth.

Case Study

Government policies: population control in China

China operates the world's most severe family planning programme. By 1954 China's population had reached 600 million and the government was worried about the pressure on food supplies and other resources. Consequently, the country's first birth control programme was introduced in 1956. This was to prove short-lived, for in 1958 the 'Great Leap

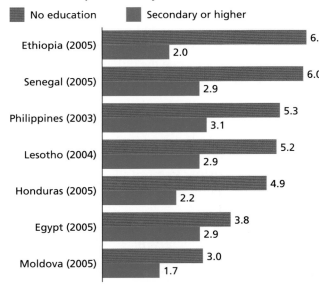

Figure 10 Fertility and level of education

Forward' began. The objective was rapid industrialisation and modernisation. The government was now concerned that progress might be hindered by labour shortages and so births were encouraged. But by 1962 the government had changed its mind, heavily influenced by a catastrophic famine due in large part to the relative neglect of agriculture during the pursuit of industrialisation. An estimated 20 million died during the famine. Thus, a new phase of birth control ensued in 1964. Just as the new programme was beginning to have some effect, a new social upheaval, the Cultural Revolution, got underway. This period, during which the birth rate peaked at 45/1000, lasted from 1966 to 1971.

With political order restored, a third family planning campaign was launched in the early 1970s with the slogan 'Late, sparse, few'. However, towards the end of the decade the government felt that its impact might falter and in 1979 the controversial 'One Child' policy was imposed. Huge economic and social pressure was placed on couples to have only one child. Some organisations, including the UN Fund for Population Activities, have praised China's policy on birth control. Many others see it as a fundamental violation of civil liberties. By 2007, the birth rate in China was down to 12/1000, the same as in the UK. This shows the huge impact the one child policy has had.

Case Study

Government policies: encouraging fertility in France

An increasing number of countries have become concerned about their falling levels of fertility. Some countries such as France have introduced clear measures to encourage couples to have more children. In France this has included:

- longer maternity and paternity leave
- higher child benefits
- improved tax allowances for larger families.

Overall, France is trying to reduce the economic cost to parents of having children. In the absence of migration, couples need to have 2.1 children if the population is to replace itself in the long term. This figure of 2.1 children is known as **replacement level fertility**. France and virtually every other country in Europe is below replacement level fertility.

Factors affecting mortality

This year, almost 11 million children under five years of age will die from causes that are largely preventable. Among them are four million babies who will not survive the first month of life. On top of that 3.3 million babies will be stillborn. At the same time, about half a million women will die in pregnancy, childbirth or soon after.

The World Health Report, 2005

In 1900 the world average for **life expectancy** is estimated to have been about 30 years but by 1950–55 it had risen to 46 years. By 1980–85 it had reached a fraction under 60 years and is presently 68 years. However, the global average masks significant differences by world region (Figure 12). The twentieth-century fall in mortality was particularly marked after the Second World War which had provided a tremendous impetus for research into tropical diseases.

> **Life expectancy at birth:** the average number of years a newborn infant can expect to live under current mortality levels.

Figure 11 Graveyard dating from the eighteenth century in the Cotswolds, UK. Inscriptions show that life expectancy in the UK at that time was very low.

The infant mortality rate is generally regarded as a prime indicator of socio-economic progress. Over the world as a whole infant mortality has declined sharply during the last half century. Between 1950 and 1955 the global average was 138/1000 but by 1975–80 it was down to 88/1000 and now it is down to 52/1000. The average for MEDCs stands at 6/1000 while the rate in LEDCs is 57/1000.

Rates of life expectancy at birth have converged significantly between rich and poor countries over the past fifty years in spite of a widening wealth gap. However, it must not be forgotten that the

Highest		Lowest	
Country	Years	Country	Years
Japan	82	Swaziland	33
Australia	81	Botswana	34
France	81	Lesotho	36
Iceland	81	Zimbabwe	37
Italy	81	Zambia	38
Sweden	81	Malawi	40
Switzerland	81	Angola	41
Austria	80	Afghanistan	42
Canada	80	Central African Republic	43
Israel	80	Mozambique	43
Malta	80		
Netherlands	80		
New Zealand	80		
Norway	80		
Singapore	80		
Spain	80		

Figure 12 Countries with the highest and lowest life expectancies, 2007

ravages of AIDS in particular have caused recent decreases in life expectancy in some countries.

The causes of death vary significantly between the more developed and less developed worlds (Figure 13). In the developing world, infectious and parasitic diseases account for over 40 per cent of all deaths. They are also a major cause of disability and social and economic upheaval. In contrast, in the more developed world these diseases have a relatively low impact. In rich countries heart disease and cancer are the big killers.

Apart from the challenges of the physical environment in many less developed countries, a range of social and economic factors contribute to the high rates of infectious diseases. These include:

- poverty
- poor access to health care
- antibiotic resistance
- evolving human migration patterns
- new infectious agents.

When people live in overcrowded and insanitary conditions, communicable diseases such as tuberculosis and cholera can spread rapidly. Limited access to healthcare and medicines means that otherwise treatable conditions such as malaria and tuberculosis are often fatal to poor people. Poor nutrition and deficient immune systems are also key risk factors for several big killers such as lower respiratory infections, tuberculosis and measles.

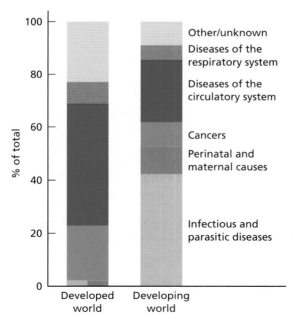

Figure 13 Contrasts in the causes of death between MEDCs and LEDCs

Factors influencing migration

These will be investigated on pages 22–26.

Activities

1 Look at Figure 7. Calculate the rate of natural change for each region.

2 Look at Figure 9. Imagine that the population of the island at the beginning of the year was 4000 rather than 5000. Calculate the rates of change for this new starting population figure.

3 Discuss three factors that cause the birth rate to vary from one part of the world to another.

4 Describe and explain the variations in life expectancy shown in Figure 12.

■ The demographic transition model

Although the populations of no two countries have changed in exactly the same way, some broad generalisations can be made about population growth since the middle of the eighteenth century. These trends are illustrated by the **demographic transition model** (Figure 14). A model is a simplification of reality, helping us to understand the most important aspects of a process.

> **Demographic transition:** the historical shift of birth and death rates from high to low levels in a population.

No country as a whole retains the characteristics of stage 1, which only applies to the most remote societies on earth such as isolated tribes in New Guinea and the Amazon Basin. All the more developed countries of the world are now in stages 4 or 5. The poorest of the less developed countries are in stage 2 but are joined in this stage by the oil-rich Middle East states where increasing affluence has not been accompanied by a significant fall in fertility. Most less developed countries which have undergone significant social and economic advances are in stage 3 while some of the newly industrialised countries such as South Korea and Taiwan have entered stage 4. Stage 5, natural decrease, is mainly confined to eastern and southern Europe at present.

The high stationary stage (stage 1): The birth rate is high and stable while the death rate is high and

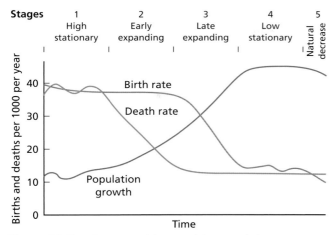

Figure 14 The demographic transition model

fluctuating due to the sporadic incidence of famine, disease and war. Population growth is very slow and there may be periods of considerable decline. Infant mortality is high and life expectancy low. A high proportion of the population is under the age of 15. Society is pre-industrial with most people living in rural areas, dependent on subsistence agriculture.

The early expanding stage (stage 2): The death rate declines to levels never before experienced. The birth rate remains at its previous level as the social norms governing fertility take time to change. As the gap between the two vital rates widens, the rate of natural change increases to a peak at the end of this stage. The infant mortality rate falls and life expectancy increases. The proportion of the population under 15 increases. The main reasons for the decline in the death rate are: better nutrition; improved public health, particularly in terms of clean water supply and efficient sewerage systems; and medical advance. Considerable **rural to urban migration** occurs during this stage.

The late expanding stage (stage 3): After a period of time social norms adjust to the lower level of mortality and the birth rate begins to decline. Urbanisation generally slows and average age increases. Life expectancy continues to increase and infant mortality to decrease. Countries in this stage usually experience lower death rates than nations in the final stage due to their relatively young population structures.

The low stationary stage (stage 4): Both birth and death rates are low. The former is generally slightly higher, fluctuating somewhat due to changing economic conditions. Population growth

is slow. Death rates rise slightly as the average age of the population increases. However, life expectancy still improves as **age-specific mortality rates** continue to fall.

The natural decrease stage (stage 5): In a limited but increasing number of countries, mainly European, the birth rate has fallen below the death rate. In the absence of net migration inflows these populations are declining. Examples of **natural decrease** include Germany, Belarus, Bulgaria and Ukraine.

Contrasts in demographic transition

There are a number of important differences in the way that less developed countries have undergone population change compared with the experiences of most more developed nations before them. In the less developed world:

- birth rates in stages 1 and 2 were generally higher
- the death rate fell much more steeply
- some countries had much larger base populations and thus the impact of high growth in stage 2 and the early part of stage 3 has been far greater
- for those countries in stage 3 the fall in fertility has also been steeper
- the relationship between population change and economic development has been more tenuous.

Activities

1 What is a *geographical model* (such as the model of demographic transition)?

2 Explain the reasons for declining mortality in stage 2.

3 Why does it take some time before fertility follows the fall in mortality?

4 Suggest why the birth rate is lower than the death rate in some countries (stage 5).

▪ Contrasting patterns of population growth

Different models of demographic transition

Although most countries followed the classical or English model of demographic transition illustrated in the last section, some countries did not. The Czech demographer Pavlik recognised two alternative types of population change, shown in Figure 15. In France the birth rate fell at about the same time as the death rate and there was no intermediate period of high natural increase. In Japan and Mexico the birth rate actually increased in stage 2 due mainly to the improved health of women in the reproductive age range.

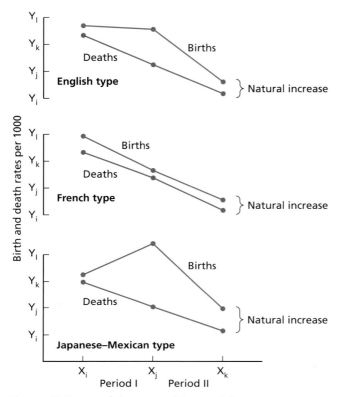

Figure 15 Types of demographic transition

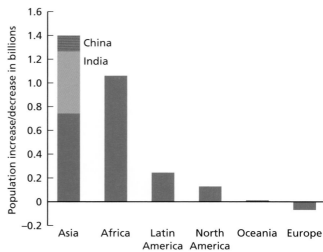

Figure 16 Projected population change by region, 2005–50

- Less than 15 per cent of the world's population lives in countries that are projected to decline in population between 2005 and 2050. These include Russia, Germany, Japan and Italy.
- Slow population growth countries will increase their populations by 25 per cent at most by 2050. China is the most important country in this group.
- Medium population growth countries include the USA, Bangladesh, Brazil and India. For example, the USA is projected to increase by 42 per cent between 2005 and 2050.
- High population growth countries accounted for only 8 per cent of world population in 2005. Except for a few oil-exporting countries, nearly all of the high population growth countries are in the UN's list of least developed countries. Many are in Africa, but the list also includes Afghanistan, Guatemala and Haiti.

The current demographic divide

Although average population growth has slowed globally, the range of demographic experience has actually widened. Growth rates have remained high in many countries while they have fallen steeply in others. These diverging trends have created a **demographic divide** between countries where population growth remains high and those with very slow growing, stagnant or declining populations. International migration is now the most unpredictable factor. For example, in the UK the birth rate has risen recently due to high levels of immigration.

Four groups of countries can be recognised in terms of projected population change to 2050 (Figure 16):

Contrasting fertility decline

Figure 17 shows contrasting fertility decline in the USA and Bangladesh. The USA is in stage 4 of demographic transition while Bangladesh is in stage 3. The graph illustrates the speed of fertility decline in LEDCs compared with the more gradual decline in MEDCs. However, fertility decline in LEDCs has also occurred at different rates (Figure 18), as it did in MEDCs in earlier years. In Figure 18, South Korea has advanced to such a high level of economic development that it is now considered an MEDC.

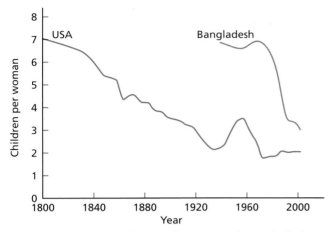

Figure 17 Fertility decline in the USA and Bangladesh, 1800–2000

Increasing mortality due to HIV/AIDS

Although in general mortality continues to fall around the world, in some countries it is rising. **HIV/AIDS** is the major reason for such increases in mortality. AIDS is the leading cause of death in Africa. One in ten people between the ages of 15 and 49 are HIV-positive in 12 countries. The majority of new infections occur among people aged 15–24. In some parts of southern Africa, 40 per cent of adults are infected.

Around 70 per cent of all people with HIV live in sub-Saharan Africa. The epidemic is particularly concentrated in the seven countries of southern Africa where the prevalence rates are 38.8 per cent in Botswana, 37.3 per cent in Swaziland, 28.9 per cent in Lesotho, 24.6 per cent in Zimbabwe, 21.5 per cent in South Africa, 21.3 per cent in Namibia and 16.5 per cent in Zambia. The factors responsible for such high rates include:

- poverty and social instability that result in family disruption
- high levels of other sexually transmitted infections
- the low status of women
- sexual violence
- high mobility, which is mainly linked to migratory labour systems
- ineffective leadership during critical periods in the epidemic's spread.

HIV/AIDS will make the populations of the 53 most affected countries 129 million lower by 2015 (a 3 per cent decrease) than they would otherwise have been. By 2050 the difference will be 480 million (an 8 per cent decrease). The seven countries worst affected are Botswana, Lesotho, Namibia, South Africa, Swaziland, Zambia and Zimbabwe. In these countries the total population is projected to be 26 million lower by 2015 (19 per cent), and 77 million lower by 2050 (36 per cent).

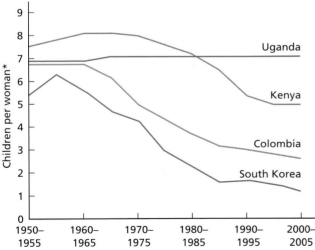

*The average total number of children a woman would have given current birth rates

Figure 18 Fertility decline in Colombia, Kenya, South Korea and Uganda

The impact of AIDS

Labour supply – the economically active population reduces as more people fall sick and are unable to work. This can have a severe impact on development. In the worst-affected countries the epidemic has already reversed many of the development achievements of recent decades. In agriculture, food security is threatened as there are fewer people able to farm and to pass on their skills.

Dependency ratio – those who contract AIDS are mainly in the economically active population. An increasing death rate in this age group increases the dependency ratio. Figure 19 (page 10) shows an estimate of the toll of AIDS on the population structure of Botswana for 2020.

Family – AIDS is impoverishing entire families and many children and old people have to take on the role of carers. Adult deaths, especially of parents, often causes households to be dissolved. The large number of orphaned children in some areas puts a considerable strain on local communities and on LEDC governments. ▶

Education – with limited investment in education many young people are still unaware about how to avoid the risk of contracting HIV. In addition there are a considerable number of teachers who have HIV/AIDS and are too ill to work. UNICEF has stressed how the loss of a significant number of teachers is a serious blow to the future development of an LEDC.

Poverty – there is a vicious cycle between HIV/AIDS and poverty. HIV/AIDS prevents development and increases the impact of poverty. Poverty worsens the HIV/AIDS situation due to economic burdens such as debt repayments and drug/medical costs.

Infant and child mortality – this increases as AIDS can be passed from mother to child.

Population structure

The structure or composition of a population is the product of the processes of fertility, mortality and migration. The most studied aspects of **population structure** are age and sex. Other aspects that can also be studied include race, language, religion, and social/occupational groups.

Population pyramids

Age and sex structure is conventionally illustrated by the use of population pyramids (Figure 21). Pyramids can be used to portray either absolute or relative data. Absolute data shows the figures in thousands or millions while relative data shows the numbers involved in percentages. The latter is most frequently used as it allows for easier comparison of countries of different population sizes. Each bar represents a five year age-group apart from the uppermost bar which usually illustrates the population 85 years old and over. The male population is represented to the left of the vertical axis with females to the right.

Demographic transition and changing population structure

Population pyramids change significantly in shape as a country progresses through demographic transition:

- The wide base of Niger's pyramid reflects extremely high fertility. The birth rate in Niger is 48/1000, one of the highest in the world. The marked decrease in width of each successive bar indicates relatively high mortality and limited life expectancy. The death rate at 15/1000 is high,

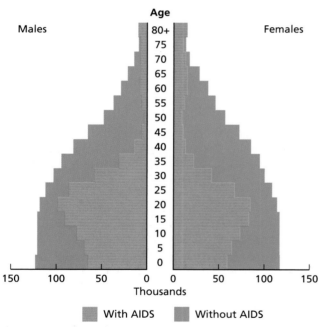

Figure 19 The impact of AIDS on the population structure of Botswana (2020 projected)

Activities

1 Describe the different types of demographic transition shown in Figure 15.

2 Describe and explain the differences in fertility decline between the USA and Bangladesh shown in Figure 17.

3 Compare fertility change in the four countries shown in Figure 18.

4 Identify the projected impact of AIDS on Botswana's population structure (Figure 19).

Figure 20 A secondary school in Morocco: the country has a youthful population

particularly considering how young the population is. The infant mortality rate is a very high 126/1000. Life expectancy in Niger is 56 years. 48 per cent of the population is under 15, with only 3 per cent 65 or more. Niger is in stage 2 of demographic transition.

- The base of the pyramid for Bangladesh is narrower, reflecting a considerable fall in fertility after decades of government-promoted birth control programmes. The fact that the 0–4 and 5–9 bars are narrower than the two bars immediately above is evidence of recent falls in

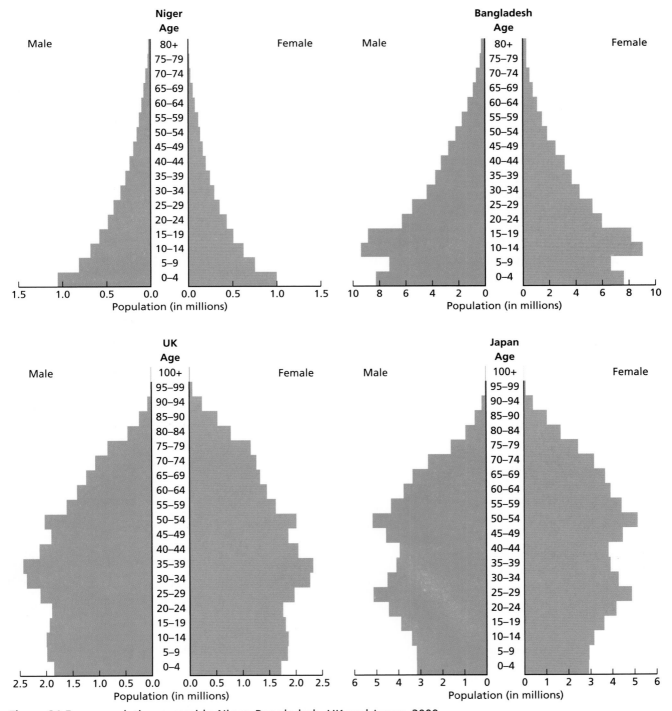

Figure 21 Four population pyramids: Niger, Bangladesh, UK and Japan, 2000

fertility. The birth rate is currently 27/1000. Falling mortality and lengthening life expectancy are reflected in the relatively wide bars in the teenage and young adult age groups. The death rate at 8/1000 is almost half that of Niger. The infant mortality rate is 65/1000. Life expectancy in Bangladesh is 62 years. 33 per cent of the population are under 15, while 4 per cent are 65 or over. Bangladesh is an example of a country in stage 3 of demographic transition.

- In the pyramid for the UK much lower fertility still is illustrated by narrowing of the base. The birth rate in the UK is only 12/1000. The reduced narrowing of each successive bar indicates a further decline in mortality and greater life expectancy compared with Bangladesh. The death rate in the UK is 10/1000, with an infant mortality rate of 4.9/1000. Life expectancy is 79 years. 18 per cent of the population are under 15, while 16 per cent are 65 or over. The UK is in stage 4 of demographic transition.

- The final pyramid (Japan) has a distinctly inverted base reflecting the lowest fertility of all four countries. The birth rate is 9/1000. The width of the rest of the pyramid is a consequence of the lowest mortality and highest life expectancy of all four countries. The death rate is 9/1000 with infant mortality at 2.8/1000. Life expectancy is 82 years. Japan has only 14 per cent of its population under 15, with 21 per cent at 65 or over. With the birth rate and the death rate in balance Japan is on the boundary of stages 4 and 5 of demographic transition.

Figure 22 provides some useful tips for understanding population pyramids. A good starting point is to divide the pyramid into three sections:

- the young dependent population
- the economically active population
- the elderly dependent population.

The dependency ratio

The **dependency ratio** is the relationship between the working or economically active population and the non-working population. The formula for calculating the dependency ratio is as follows:

The dependency ratio

$$= \frac{\% \text{ pop aged } 0\text{--}14 + \% \text{ pop aged } 65+}{\% \text{ pop } 15\text{--}64}$$

The dependency ratio for more developed countries is usually between 50 and 75. In contrast, less developed countries typically have ratios between 85 and 105.

The **youth dependency ratio** is the ratio of the number of people under 15 to those 15–64 years of age. The **elderly dependency ratio** is the ratio of the number of people over 64 years to those 15–64 years of age.

Sex structure

The **sex ratio** is the number of males per 100 females in a population. Male births consistently exceed female births due to a combination of biological and social reasons. For example, in terms

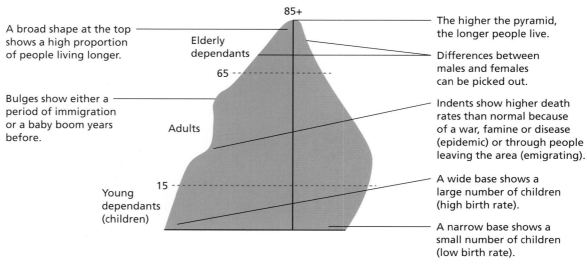

Figure 22 An annotated population pyramid

of the latter, more couples decide to complete their family on the birth of a boy than on the birth of a girl. In the UK 105 boys are born for every 100 girls. However, after birth the gap generally begins to narrow until eventually females outnumber males, as at every age male mortality is higher than female mortality. This process happens most rapidly in the poorest countries where infant mortality is markedly higher among males than females. Here the gap may be closed in less than a year. In the UK it is not until the 45–59 age group that females outnumber males. In the age group 85 and over females make up 74 per cent of the population.

However, there are anomalies to this picture. In countries where the position of women is markedly subordinate and deprived, the overall sex ratio may show an excess of males. Such countries often exhibit high mortality rates in childbirth. For example, in India there are 107 males per 100 females for the population as a whole.

A report published in China in 2002 recorded 116 male births for every 100 female births due to the significant number of female fetuses aborted by parents intent on having a male child. Even within countries there can be significant differences in the sex ratio. In the USA, Alaska has the highest ratio at 103.2, while Mississippi has the lowest at 92.2.

Population structure: differences within countries

In countries where there is strong rural to urban migration, the population structures of the areas affected can be markedly different (Figure 23). These differences show up clearly on population pyramids. Out-migration from rural areas is age-selective, with single young adults and young adults with children dominating this process. Thus the bars for these age groups in rural areas affected by out-migration will indicate fewer people than expected in these age groups.

In contrast, population pyramids for urban areas attracting migrants will show age-selective in-migration, with substantially more people in these age groups than expected. Such migrations may also be sex-selective. If this is the case it should be apparent on the population pyramids.

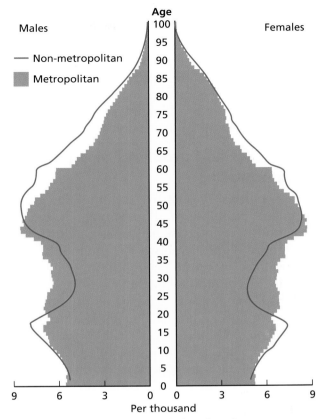

Figure 23 Population pyramid showing the contrast between rural and urban populations in Canada

Activities

1 a) Describe and explain the differences between the four population pyramids shown in Figure 21.
  b) Produce a table to show all the statistical data given for the four countries. Keep the same order of countries as in the text. For how many of the data sets is there a clear trend?

2 Using the data in Figure 24, calculate the dependency ratios for Nigeria and the UK.

3 How and why might the population structures of rural and urban areas in the same country differ?

4 a) Define the term *sex ratio*.
  b) Suggest reasons why the sex ratio can differ both between and within countries.

Country	% population under 15	% population 65 and over
Nigeria	45	3
UK	18	16

Figure 24 A population comparison

Consequences of different patterns of population growth

Different patterns of population growth can bring both benefits and problems to the countries concerned. This is particularly the case when countries have a very high percentage of either young or old people in their populations.

The benefits and problems of a large young population

Rapid population growth results in a large young dependent population. The young dependent population is defined as the population under 15 years of age. Figure 25 shows the variation around the world average of 28 per cent. The 41 per cent for Africa is over two and a half times higher than the figure for Europe. The highest figures for individual countries are in Uganda (50 per cent), Guinea-Bissau, Mali, and Niger (all 48 per cent).

Countries with large young populations have to allocate a substantial proportion of their national resources to look after them. Young people require resources for health, education, food, water and housing above all. The money required to cover such needs may mean there is little left to invest in agriculture, industry and other aspects of the economy. A government of a less developed country might see this as being too large a demand on the country's resources and as a result may introduce family planning policies to reduce the birth rate.

However, individual parents may have a different view, where they see a large family as valuable in terms of the work children can do on the land. Alongside this, people in poor countries often have to rely on their children in old age because of the lack of state welfare benefits.

As a large young population moves up the age ladder over time, it will provide a large working population when it enters the **economically active age group** (15–64). This will be an advantage if a country can attract sufficient investment to create enough jobs for a large working population. Then, the large working population will contribute money in taxes to the country which can be invested in many different ways to improve the quality of life and to attract more foreign investment. Such a situation can create an upward spiral of economic growth.

World	28
Africa	41
North America	20
Latin America/Caribbean	30
Asia	28
Europe	16
Oceania	25

Figure 25 The percentage of total population under 15 years of age, 2007

On the other hand, if there are few employment opportunities for a large working population, the unemployment rate will be high. The government and most individuals will have little money to spend and the quality of life will be low. Many young adults may seek to emigrate because of the lack of opportunities in their own country.

Eventually, the large number of people in this age group will reach old age. If most of them enter old age in poverty, this creates even more problems for the government.

The benefits and problems of an ageing population

The world's population is ageing significantly. Ageing of population is a rise in the **median age** of a population. It occurs when fertility declines while life expectancy remains constant or increases.

The following factors have been highlighted by the United Nations:

- The global average for life expectancy increased from 46 years in 1950 to nearly 65 in 2000. It is projected to reach 74 years by 2050.
- In LEDCs the population aged 60 years and over is expected to quadruple between 2000 and 2050.
- In MEDCs the number of older people was greater than that of children for the first time in 1998. By 2050 older people in MEDCs will outnumber children by more than two to one.
- The population aged 80 years and over (the oldest old) numbered 69 million in 2000. This was the fastest-growing section of the global population which is projected to increase to 375 million by 2050.
- Europe is the 'oldest' region in the world. Those aged 60 years and over currently form 20 per cent of the population. This should rise to 35 per cent by 2050.

- Japan is the oldest nation with a median age of 41.3 years, followed by Italy, Switzerland, Germany and Sweden.
- Africa is the 'youngest' region in the world, with the proportion of children accounting for 43 per cent of the population today. However, this is expected to decline to 28 per cent by 2050. In contrast the proportion of older people is projected to increase from 5 per cent to 10 per cent over the same time period.
- There are already over 50 nations with total fertility rates at or below 2.1 (replacement level). By 2016 the UN predicts there will be 88 nations in this category.

Figure 26 shows that 7 per cent of the world's population are aged 65 years and over. On a continental scale this varies from only 3 per cent in Africa to 16 per cent in Europe. Population projections show that the world population 65 years and over will rise to 10 per cent in 2025, and to 16 per cent by 2050.

World	7
Africa	3
North America	12
Latin America/Caribbean	6
Asia	6
Europe	16
Oceania	10

Figure 26 The percentage of total population 65 years and over, 2007

The problem of **demographic ageing** has been a concern of MEDCs for some time, but it is now also beginning to alarm LEDCs. Although ageing has begun later in LEDCs, it is progressing at a faster rate. This follows the pattern of previous demographic change such as declining mortality and falling fertility where change in LEDCs was much faster than that previously experienced by MEDCs. For example, Figure 27 shows the projected decline in the **support ratio** for the world as a whole and a number of individual countries. The support ratio measures the relationship between the numbers of working and retired people. It is often used as an alternative to the dependency ratio.

Demographic ageing will put healthcare systems, public pensions, and government budgets in general, under increasing pressure. Four per cent of

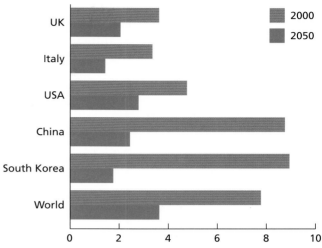

Figure 27 Support ratio forecasts

the USA's population was 65 years of age and older in 1900. By 1995 this had risen to 12.8 per cent and by 2030 it is likely that one in five Americans will be senior citizens. The fastest growing segment of the population is the so-called 'oldest-old': those of 80 years or more. It is this age group that is most likely to need expensive residential care. The situation is similar in other developed countries.

Some countries have made relatively good pension provision by investing wisely over a long period of time. However, others have more or less adopted a pay-as-you-go system, as the elderly dependent population rises. It is this latter group who will be faced with the biggest problems in the future.

For much of the post-1950 period the main demographic problem has been generally perceived as the 'population explosion', a result of very high

Figure 28 A deserted village in northern Spain due to out-migration and population ageing

fertility in the developing world. However, greater concern is now being expressed about demographic ageing in many countries, where difficult decisions about the reallocation of resources are having to be made. Very few countries are generous in looking after their elderly at present. Poverty amongst the elderly is a considerable problem but technological advance might provide a solution by improving living standards for everyone. If not, other less popular solutions, such as increased taxation, will have to be examined.

Activities

1 Draw a graph to illustrate the data in Figure 25.

2 Discuss the benefits and problems of a young dependent population.

3 Describe the variations shown in Figure 27.

4 Why is a large elderly dependent population generally viewed as a problem?

5 Discuss the possible benefits of a large elderly population.

■ Population growth and resources

The population and resource relationship

The relationship between population and resources has concerned people for thousands of years. For example, Confucius, the ancient Chinese philosopher, said that excessive population growth reduced output per worker and caused living standards to fall.

The Revd Thomas Malthus (1766–1834) was concerned that population would grow faster than the supply of food. Malthus thought that an increased food supply was achieved mainly by bringing more land into arable production. He maintained that while the supply of food could, at best, only be increased by a constant amount in arithmetical progression (1 - 2 - 3 - 4 - 5 - 6), the human population tends to increase in geometrical progression (1 - 2 - 4 - 8 - 16 - 32), multiplying itself by a constant amount each time. In time, population would outstrip food supply until a catastrophe occurred. This would be in the form of famine, disease or war. Such catastrophes would occur as human groups fought over increasingly

scarce resources. These limiting factors maintained a balance between population and resources in the long term.

Clearly Malthus was influenced by events in and before the eighteenth century and could not have foreseen the great technological advances that were to unfold in the following two centuries which have allowed population to grow at unprecedented rates alongside a huge rise in the exploitation and use of resources. Figure 29 shows the population–technology–resources relationship.

Today the pessimistic lobby who fear that population growth will outstrip resources leading to the consequences predicted by Thomas Malthus are referred to as Malthusians or neo-Malthusians. The anti-Malthusians are optimists who argue that either population growth will slow well before the limits of resources are reached or that the ingenuity of humankind will solve resource problems when they arise. For example, in 1965 Esther Boserup argued that technological advance would ensure that food supply kept up with population. Figure 30 sums up these opposing views.

Underpopulation, overpopulation and optimum population

The idea of **optimum population** has been mainly understood in an economic sense (Figure 32). At first, an increasing population allows for a fuller exploitation of a country's resource base causing living standards to rise. However, beyond a certain level rising numbers place increasing pressure on resources and living standards begin to decline. The highest average living standards mark the optimum population. Before that population is reached the country or region is said to be **underpopulated**. As the population rises beyond the optimum the country or region is said to be **overpopulated**.

Where an individual country is placed in terms of its relationship between population and resources is a matter of opinion. There may be a big difference

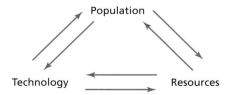

Figure 29 The population, resources and technology relationship

There are two opposing views of the effect of population growth:

1 Neo-Malthusian

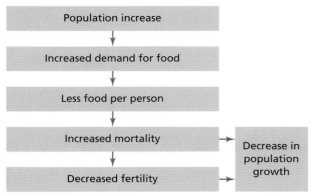

Expanding population means increasing food production causing environmental and financial problems.

2 Resource optimists (e.g. Boserup)

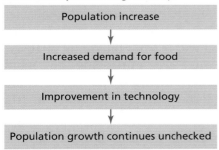

People are the ultimate resource – through innovation or intensification humans can respond to increased numbers.

Figure 30 The opposing views of Malthus and Boserup

Figure 31 A major irrigation canal in the Nile valley. The construction of the irrigation system greatly increased the carrying capacity of the land.

in the views of people living in the same country. Such views can change over time, particularly with economic cycles.

The Netherlands and the UK are two of the most densely populated countries in Europe. Not everyone in these countries thinks they are overpopulated, but it does seem that an increasing number of people are of this opinion. In the UK, an organisation called the Optimum Population Trust states that 30 million is the optimum population for the country. At present the population of the UK is about 60 million. Signs of population pressure in both countries include:

- intense competition for land
- heavy traffic congestion
- high house prices
- high environmental impact of economic activity
- pressure on water resources.

Two of the most sparsely populated developed countries in the world are Australia and Canada. Throughout the history of both countries the general view has been that they would benefit from higher populations. Thus Australia and Canada have welcomed significant numbers of immigrants. However, in recent years, with an uncertain economic climate, both countries have been much

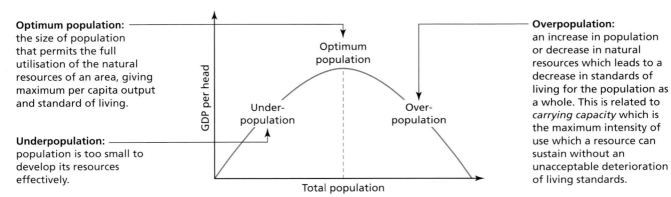

Figure 32 Optimum population, overpopulation and underpopulation

more selective in terms of immigration. Although both countries are very large in size, they have large areas of inhospitable landscape.

In the developing world, China and Bangladesh are countries that many view as overpopulated. The 'one child' policy confirms the Chinese government view. Bangladesh has one of the highest population densities in the world and struggles to provide for many in its population. In contrast, Malaysia has been concerned that it was underpopulated. In 1982, when the country's population was below 15 million, the government announced that the country should aim for an ultimate population of 70 million. A range of inducements were put in place to encourage people to have larger families.

Global resources

There are many indications that human population is pushing up against the limits of the Earth's resources. For example:

- Twenty-five per cent of children worldwide suffer from protein-energy malnutrition.
- The long-term trend for grain production per person is falling.
- About 40 per cent of agricultural land is moderately degraded and 9 per cent is highly degraded.
- Water scarcity already affects every continent and four of every ten people in the world.
- A quarter of all fish stocks are overharvested.
- There are concerns that global peak oil production will come as early as the next decade.

The new global food problem

In December 2007 the price of wheat broke through the $10 a bushel level, sparking protests in Pakistan and other countries as rising wheat prices were passed on to consumers. There can be little doubt that the days of easy grain surpluses are a thing of the past. For example, India, the second largest consumer of wheat, became a large net importer in 2006 after a six-year period as a net exporter. At about the same time Russia was considering curbs on wheat exports to prevent domestic prices of this cereal rising too rapidly. In addition, Australia, the third biggest exporter of wheat, warned that its output might be 18 per cent less than a previous government estimate due to a second year of drought. A late season drought in

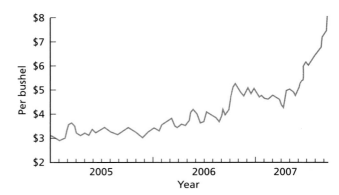

Figure 33 Global wheat price increases

the Ukraine had a considerable impact on production. In Morocco the crop was down 76 per cent. Argentina temporarily halted wheat exports to assess damage caused by cold weather.

Demand for grain and other crops is rising:

- Global population continues to increase significantly and will reach 7 billion by 2010.
- Living standards are improving in many countries, especially in highly populated China and India. Higher incomes result in the increasing demand for meat. However, it takes 7 kg of grain to produce 1 kg of beef. Consumers in NICs are following the lifestyles developed in MEDCs.
- Agricultural resources are being diverted from food production to biofuel manufacture because of concerns about energy security. This is reducing food production significantly in some areas.

The following are the main problems with the supply of grain:

- Most good-quality farmland is already being used.
- About a third of this has been significantly degraded by intensive farming in the last half-century.
- The world's deserts expanded by 160 million hectares between 1970 and 2000.
- The global land area used for the cultivation of wheat and barley has been falling for 25 years.
- Drought and other adverse environmental factors have significantly reduced yields in key producing countries. More and more countries are becoming concerned about the impact of climate variability on food production.

Figure 34 Low population density on the west coast of Ireland

Figure 35 High population density – Cairo, Egypt

■ Influences on population distribution and density

Population density is the average number of people per square kilometre ($/km^2$) in a country or region. **Population distribution** is the way that the population is spread out over a given area, from a small region to the Earth as a whole. Figure 36 shows the global distribution of population using a dot map. Areas with a high population density are said to be **densely populated**. Regions with a low population density are **sparsely populated**.

Figure 37 (page 20) shows the density of population by world region. The huge overall contrast between the more developed and less

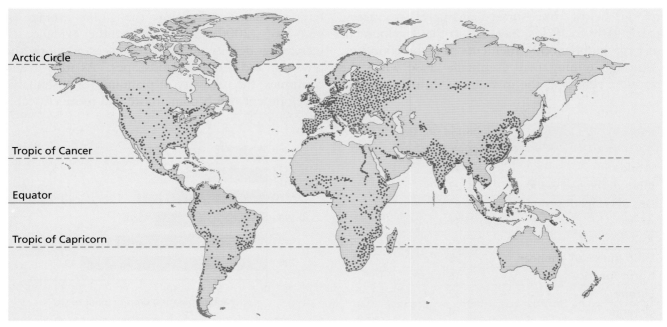

Figure 36 Dot map showing world population distribution

developed worlds is very clear. The average density in the less developed world is more than double that of the more developed world. North America (15/km²) and Oceania (4/km²) have the lowest population densities of all the world regions. However, the overall difference between the developed and developing worlds is largely accounted for by the extremely high figure for Asia (126/km²).

The average density figure for each region masks considerable variations. The most uniform distributions of population occur where there is little variation in the physical and human environments. Steep contrasts in these environments are sharply reflected in population density. People have always avoided hostile environments if a reasonable choice has been available. Look at an atlas map of the world illustrating population density. Now look at world maps of relief, temperature, precipitation and vegetation. See the low densities associated with high altitudes, polar regions, deserts and rainforests. More detailed maps can show the influence of other physical factors such as soil fertility, natural water supply and mineral resources.

Areas of low soil fertility have been avoided from the earliest times of settlement as people looked for more productive areas in which to settle. Water supply has always been vitally important. This is why so many settlements are historically located by a) rivers b) lakes c) springs and d) where artesian wells could be dug to access aquifers (water-bearing rocks) below the surface. Mineral resources, particularly coalfields, have led to the development of large numbers of settlements in many countries. Although mining may eventually cease when the resource runs out, the investment in infrastructure (housing, railways, roads, etc.) over time usually means that the settlement will continue. The Ruhr coalfield in western Germany, once the most productive in Europe, now has only a few working mines remaining. However, it is still one of the most densely populated regions of Europe. But mining settlements in very hostile environments such as Alaska, Siberia and the Sahara Desert may be abandoned when mining stops.

The more advanced a country is the more important the elements of human infrastructure become in influencing population density and distribution. While a combination of physical factors will have decided the initial location of the major urban areas, once such towns or cities reach a certain size, economies of scale and the process of cumulative causation ensure further growth. As a country advances, the importance of agriculture decreases and employment relies more and more on the secondary and tertiary sectors of the economy which are largely urban based. The lines of communication and infrastructure between major urban centres provide opportunities for further urban and industrial location.

Population growth and increasing density

Over the last 50 years average population density in Brazil has increased considerably (Figure 38). Brazil's population density is now more than three times what it was in 1950. However, the current figure remains well below the global average of 45.6/km² (2000). Nevertheless, the evidence of increasing density in Brazil is clear in both the physical and human landscapes. As this trend continues, the necessity of planning for sustainable development will become more and more critical.

Region	Population density (people/km²)
World	49
More developed world	27
Less developed world	65
Africa	31
Asia	126
Latin America/Caribbean	28
North America	15
Europe	32
Oceania	4

Figure 37 Variations in world population density

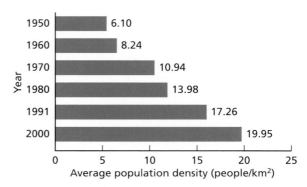

Figure 38 Change in average population density in Brazil, 1950–2000

Case Study

Population distribution and density in North America

North America has a low population density compared with most other parts of the world. The USA has an average of 31 persons/km², while Canada has only 3/km². In both countries population is highly concentrated in some areas while large expanses of land elsewhere are very sparsely settled (Figure 39). Very few people live in the cold, dry and mountainous regions. The influence of low temperature is very clear in the north and largely explains why 75 per cent of Canadians live within 160 km of the main border with the USA. Life is extremely difficult in the permafrost environment of the northlands and, apart from the native Inuit and Indians, the few people living there are mainly involved in the exploitation of raw materials and in maintaining defence installations, although the role of tourism is expanding.

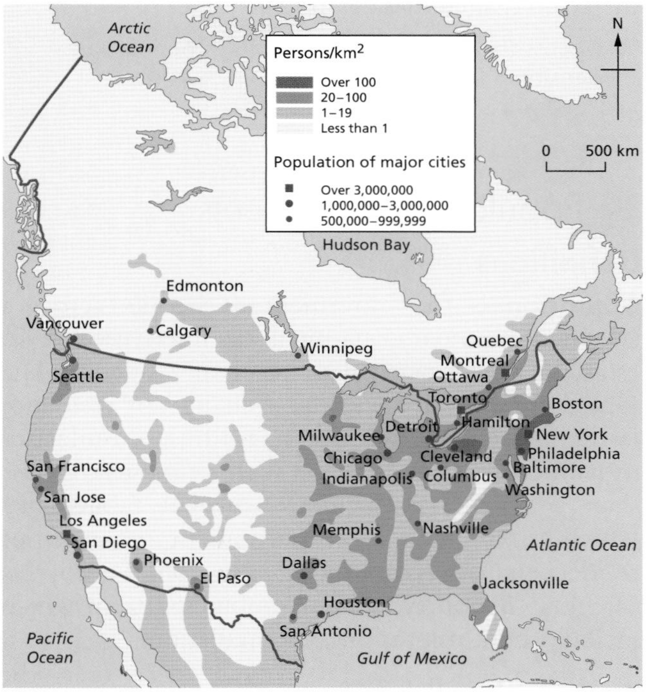

The mountain ranges of the west form imposing landscapes but economic opportunities are scarce in this rugged environment. For example, Cheyenne (altitude 1850 m), the largest settlement in Wyoming, has a population of only 50,000. The Appalachian mountains in the east are both lower in height and less extensive in area; nevertheless, the most isolated areas are very sparsely peopled.

Although much of the south-western USA is desert or semi-desert, no country in the world has been more successful in watering its drylands. Expensive irrigation schemes have opened up many parts of the region to farming, settlement and industry. Cities such as Phoenix, Tucson and Las Vegas, standing out like oases in the desert, are clear evidence of the very high level of investment. Yet large parts of the south-west still remain empty and may well continue to be in the future as the water supply problem in the region has now reached crisis point.

Figure 39 Population density of North America

Figure 40 Low population density – Interstate 80, Nevada

Figure 41 High population density – apartment blocks, Waterfront, Vancouver

In the USA the greatest concentration of population is in the north-east, the first region of substantial European settlement. By the end of the nineteenth century it had become the greatest manufacturing region in the world and by the 1960s the highly urbanised area between Boston and Washington had reached the level of a **megalopolis**. Although other parts of the country are growing at a faster rate, the intense concentration of job opportunities in the north-east will ensure that it remains the most densely populated part of the continent in the foreseeable future.

The coastline and major lowland valleys are as attractive to settlement now as they have been in the past. More than half of all Americans live in the counties (sub-divisions of states) adjacent to the Atlantic and Pacific Oceans, the Gulf of Mexico and the Great Lakes. The fastest coastal growth has been along the Pacific and Gulf of Mexico coasts where the population has doubled since 1960.

Soil fertility is another influence, but the high level of mechanisation in modern agriculture means that this factor was more important in the past. The location of other natural resources such as coal, iron ore and oil has also been an attraction to settlement, but again such influence has lessened. Most North Americans today are employed in urban-based service or manufacturing jobs and the availability of employment in these sectors is the most important influence on the distribution and density of population today.

The main concentration of population in Canada is in the southern parts of Ontario and Quebec. This region has a combination of physical and human advantages greater than that found in most other parts of the country. Two of Canada's three 'million-size' cities, Toronto and Montreal, are located here. The third-ranking city, Vancouver, is situated on the Pacific coast. A large area of low to moderate density is found in the cereal farming region of the Prairie Provinces (Manitoba, Saskatchewan, Alberta).

Population density is not a static phenomenon. The westward movement of population in North America, which began within a century of the initial establishment of settlement, has continued unabated in the twentieth century in both the USA and Canada, while the former has also experienced rapid growth in many southern states. In 1980 the mean centre of population in the USA crossed the Mississippi River for the first time and for the first time in US history more Americans lived in the south and west than in the north-east and mid-west.

Activities

1 Which global environments have the lowest population densities? Explain why.

2 Which global environments have the highest population densities? Explain why.

3 How can the population density of a country or region change considerably over time?

4 Discuss the reasons for variations in population density in the USA.

■ Population movements

Definitions

Migration is the movement of people across a specified boundary, national or international, to establish a new permanent place of residence. The UN defines 'permanent' as a change of residence lasting more than one year. In contrast, **circulatory movements** are movements with a timescale of less than a year. This includes seasonal movements which involve a semi-permanent change of residence. Daily commuting also comes into this category. **Mobility** is an all-embracing term which includes both migration and circulation.

Push and pull factors

Figure 42 shows the main push and pull factors relating to migration. **Push factors** are negative conditions at the point of origin which encourage or force people to move. **Pull factors** are positive conditions at the point of destination which encourage people to migrate. The nature of push and pull factors varies from country to country (and from person to person) and changes over time.

Voluntary and forced migrations

Figure 43 shows the main types of migrations and the barriers that potential migrants can face. Today, immigration laws present the greatest obstacle to most potential international migrants whereas in the past the physical dangers encountered on the journey often presented the greatest difficulty.

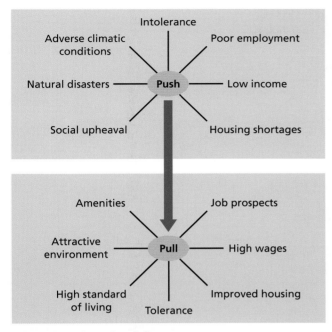

Figure 42 Push and pull factors

The big distinction is between **voluntary migration** and **forced migration**. In voluntary migration the individual has a free choice about whether to migrate or not. In forced migrations, people are made to move against their will. The abduction and transport of Africans to the Americas as slaves was the largest forced migration in history. In the seventeenth and eighteenth centuries 15 million people were shipped across the Atlantic Ocean as slaves. The expulsion of Asians from Uganda in the 1970s when the country was under the dictatorship of Idi Amin, and the forcible movement of people from parts of the former Yugoslavia under the policy of 'ethnic cleansing', are much more recent examples. Migrations may also be forced by natural disasters (volcanic eruptions, floods, drought, etc.) or by environmental catastrophe such as nuclear contamination in Chernobyl.

Figure 44 (page 24), illustrating the world's major international migration flows, shows that every world region is affected.

Migration trends

One in every 35 people around the world is living outside the country of their birth. This amounts to about 175 million people, more than ever before. Foreign-born populations are rising in both more developed and less developed countries (Figure 45, page 24). Recent migration data shows these facts:

- With the growth in the importance of labour-related migration and international student mobility, migration has become increasingly temporary and circular in nature. For example, in 2001 there were 475,000 foreign students in the USA. The international mobility of highly skilled workers increased substantially in the 1990s.
- The spatial impact of migration has spread with an increasing number of countries affected as either points of origin or destination. While many traditional migration streams remained strong, significant new streams have developed.

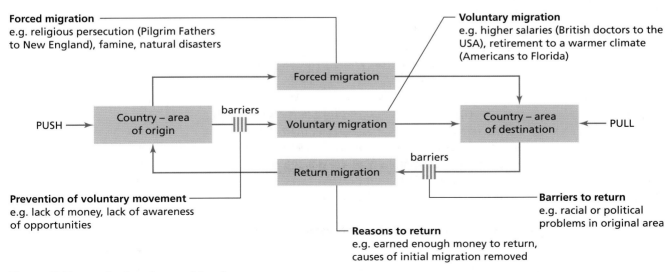

Figure 43 Types of migration and barriers

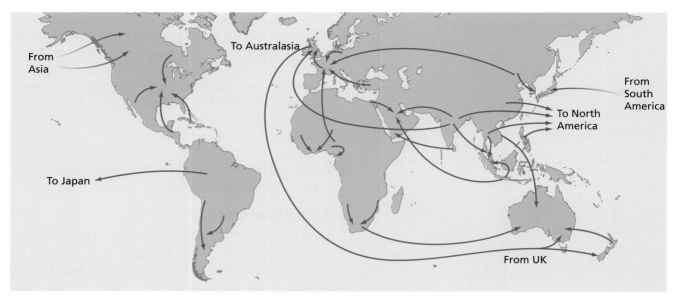

Figure 44 Major international migration flows

- The proportion of female migrants has steadily increased (now over 47 per cent of all migrants). For some countries of origin, women now make up the majority of contract workers (for example the Philippines, Sri Lanka, Thailand and Indonesia).
- The great majority of international migrants move from LEDCs to MEDCs. However, there are also strong migration links between some LEDCs, in particular between low- and middle-income countries.
- MEDCs have reinforced controls, in part in response to security issues, but also to combat illegal immigration and networks that deal in trafficking and exploitation of human beings.

Figure 46 Southall, the centre of London's Indian community

Globalisation in all its aspects has led to an increased awareness of opportunities in other countries. With advances in transportation and communication and a reduction in the real cost of both, the world's population has never had a higher level of potential mobility. Also, in various ways, economic and social development has made people more mobile and created the conditions for emigration.

Rising remittance payments

Remittances are a major economic development factor in LEDCs. In 2003, remittances from migrants working abroad reached an estimated $100 billion, an increase of about 15 per cent over the previous year. Some economists argue that remittances are the developing world's most effective source of financing:

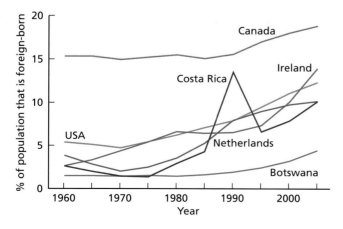

Figure 45 Foreign-born populations

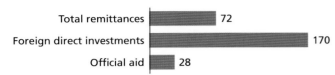

Figure 47 Financial flows to LEDCs (billion dollars), 2001

- Although foreign direct investment is larger it varies with global economic fluctuations.
- Remittances exceed considerably the amount of official aid received by LEDCs (Figure 47).

Remittances can account for up to 15 per cent of annual GDP in LEDCs. They have been described as 'globalisation bottom up'. Migration advocates stress that these revenue flows:

- help alleviate poverty
- spur investment
- cushion the impact of global recession when private capital flows decrease.

The major sources of remittances are the USA, Western Europe and the Gulf (Figure 48). The number of foreigners working in these areas is rising significantly. About 1.3 million migrants settle in the USA annually, around one-third of them illegally. The top destinations of remittances are India, Mexico and the Philippines. The 20 million people who make up the Indian **diaspora** are spread over 135 countries. In 2003 they sent back to India almost $15 billion – a source of foreign exchange that exceeds revenues generated by India's software industry. The Indian state of Kerala has nearly a million 'Gulf wives' living apart from their husbands.

Apart from the money that migrants sent directly to their families, their home communities and countries also benefit from:

- donations by migrants to community projects
- the purchase of goods and services produced in the home country by migrants working abroad
- increased foreign exchange reserves.

All three forms of economic benefit mentioned above combine to form a positive **multiplier effect** in donor countries.

The pro-migration agenda of developing nations

Many developing countries are looking to MEDCs to adopt a more favourable attitude to international migration. These are their arguments:

Top sources

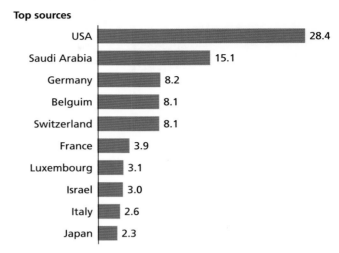

Top destinations

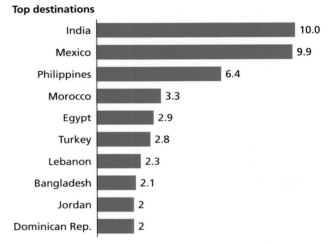

Figure 48 Sources and destinations of remittances (billion dollars), 2001

- Emigration can significantly ease the pressure on unemployment rates in developing countries as well as bringing in much needed remittance money.
- Labour mobility results in higher global standards of living. According to a recent World Bank report, a 10 per cent increase in emigration from a developing country will produce a 2 per cent decline in the number of people living on less than one dollar a day.
- Remittance money can have important trickle-down benefits. Edward Turner (University of California) has quantified the multiplier effect of remittances from the USA to Mexico. He calculates that for every dollar sent home to Mexico, three dollars more is generated in the form of construction material, food or contract work.

Figure 49 Chinatown, San Francisco: a major Chinese community

Internal population movements

Population movement within countries is at a much higher level than movements between countries. In both MEDCs and LEDCs significant movements of people take place from poorer regions to richer regions as people seek employment and higher standards of living. In LEDCs, much of this migration is in fact from rural to urban areas.

Rural to urban migration

In Brazil, there has been a large migration from the poor north-east region to the more affluent south-east. Within the north-east movement from rural areas is greatest in the Sertao, the dry interior which suffers intensely from unreliable rainfall. However, poor living standards and a general lack of opportunity in the cities of the north-east has also been a powerful incentive to move. Explaining the attraction of urban areas in the south-east demands more that the 'bright lights' scenario that is still sometimes used. The Todaro model presents a more realistic explanation. According to this model, migrants are all too well aware that they may not find employment by moving to, say, São Paulo. However, they calculate that the probability of employment, and other factors that are important to the quality of life of the individual and the family, is greater in the preferred destination than at their point of origin.

The largest rural–urban migration in history is now taking place in China. Large-scale rural–urban migration really began in the late 1980s when industrial development in the coastal cities began to take off.

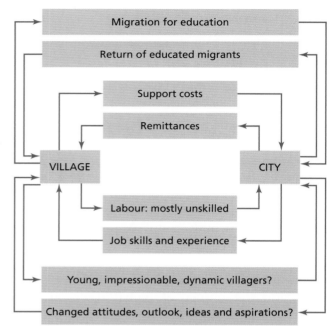

Figure 50 Costs and returns of migration

Figure 50 provides a useful framework for understanding the costs and returns from migration between rural and urban areas. It highlights the main factors which determine how rural areas are affected by migration – namely the two-way transfers of labour, money, skills and attitudes.

Counterurbanisation

In MEDCs, two significant trends can be identified concerning the redistribution of population since the late eighteenth century. The first, urbanisation, lasted until about 1970, while the second, **counterurbanisation**, has been dominant since that time. The process of urbanisation had a considerable impact on many rural areas where depopulation occurred because of it. Counterurbanisation, which has resulted in a renaissance in the demographic fortunes of rural areas, is often referred to as the 'population turnaround'. There has been considerable debate as to whether this trend will be long-term or relatively short-lived.

Activities

1 Discuss three significant push factors.

2 Suggest how the barriers to migration have changed over time.

3 Describe the data shown in Figures 47 and 48.

Types of settlements

Most of us live in settlements, and most of us take them for granted. And yet there is a huge variety of settlements, and they are changing rapidly. For example, some settlements in rural areas differ greatly from those in urban areas, although the distinction between them is becoming less clear. In developing countries large cities are growing at the expense of rural areas, despite a recent movement out of some very large cities or 'supercities'. Population change, technological developments and changing lifestyles are having a tremendous impact on settlement geography.

In this unit we look at the size, development and function of rural and urban settlements. We begin with rural settlements and examine their pattern, site and situation, function, and hierarchy. We examine the characteristics of land use in MEDCs and LEDCs, describe the problems of urban areas in MEDCs and LEDCs, and consider possible solutions to these problems. We also look at the impact on the environment as a result of urbanisation and examine possible solutions to reduce their impacts.

■ Rural settlement

A settlement is defined as a place in which people live and where they carry out a variety of activities, such as residence, trade, agriculture and manufacturing. Most rural settlements are hamlets and villages, although not all are. The study of rural settlement includes:

- pattern
- form (or shape)
- site and situation
- function and hierarchy
- change.

Pattern

A **dispersed** settlement pattern is one in which individual houses and farms are widely scattered throughout the countryside (Figure 1). It occurs when farms or houses are set among their fields or spread along roads, rather than concentrated on one point. They are common in sparsely populated areas,

Figure 1 Dispersed settlement

such as the Australian outback and the Sahel region of Africa, and in recently settled areas, such as after the creation of the Dutch polders. The enclosure of large areas of common grazing land, into smaller fields separated by hedges, led to a dispersed settlement pattern. This happened because it became more convenient to build farmhouses out in the fields of the newly established farms. Similarly, the break-up of large estates (such as after the Reformation) also led to a dispersed settlement pattern. In areas where the physical geography is quite extreme (too hot or cold, wet or dry) there is likely to be a low population density, and a poor transport network, which discourages settlement.

Nucleated settlements are those in which houses and buildings are tightly clustered around a central feature such as a church, village green or crossroads (Figure 2, page 28). Some may be linear (see page 28). Very few houses are found in the surrounding fields. Such nucleated settlements are usually termed hamlets or villages according to their size and/or function. A number of factors favour nucleation:

- joint and co-operative working of the land – people live in nearby settlements
- defence, for example hilltop locations, sites within a meander or within walled cities such as Jericho
- shortage of water causing people to locate in areas close to springs

Figure 2 Nucleated settlement Mgwali, South Africa

Figure 4 Malaysian longhouse

- swampy conditions which force people to locate settlements on dry ground
- near important junctions and crossroads as these favour trade and communications
- in some countries, the government has encouraged people to live in nucleated settlements, such as the Ujaama scheme in Tanzania, kibbutzim in Israel and communes in China.

Village form

Village form refers to shape (Figure 3). In a **linear** settlement houses are spread out along a road or a river. This suggests the importance of trade and transport during the growth of the village. Linear villages are also found where poor drainage prohibits growth in a certain direction. In the rainforests of Sarawak (Malaysia), many of the longhouses are located alongside rivers (Figure 4).

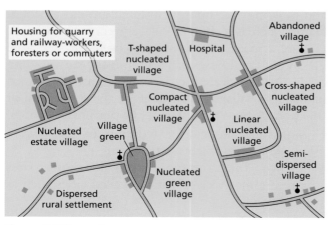

Figure 3 Village shapes

Cruciform settlements are found at intersections of roads and usually consist of lines of buildings radiating out from the crossroads. The exact shape depends on the position of the roads and the amount of infilling that has since taken place. By contrast, a **green village** consists of dwellings and other buildings, such as a church, clustered around a small village green or common, or other open space. In South Africa **ring villages** are formed where the houses, called kraals, are built around an open area.

Site and situation

The **site** of a settlement is the actual land on which a settlement is built, whereas the **situation** (or position) is the relationship between a particular settlement and its surrounding area. In the past geographers have emphasised the importance of physical conditions in the pattern of settlement, land tenure and type of agriculture practised. Increasingly, social and economic factors are important, especially in explaining recent changes in rural settlements.

Early settlers took into account advantages and disadvantages of alternative sites for agriculture and housing. These included:

- availability of water – necessary for drinking, cooking, washing, as a source of food supply and transport
- freedom from flooding – but close to the flooded areas as these are rich in fertile river deposits
- level sites to build on – but they are less easy to defend
- local timber for construction and fuel

- sunny south-facing slopes (in the northern hemisphere) as these are warmer than north-facing slopes and therefore better for crop growth
- proximity to rich soils for cultivation and lush pasture for grazing
- the potential for trade and commerce, such as close to bridges or weirs, near confluence sites, head of estuaries, point of navigation and at an upland gap.

A **dry point** is an elevated site in an area of otherwise poor natural drainage. It includes small hills (knolls) and islands. Gravel terraces along major rivers are well favoured. Water supply and fertile alluvial soils and the use of a valley as a line of communication are all positive advantages.

A **wet point** is a site with a reliable supply of water from springs or wells in an otherwise dry area. Spring line villages at the foot of chalk and limestone ridges are good examples. **Spring line settlements** occur when there is a line of sites where water is available.

Some hilltop villages suggest that the site was chosen to avoid flooding in a marshy area as well as for defence. Villages at important river crossings are excellent centres of communications.

■ Settlement hierarchy

The term **hierarchy** means 'order'. Settlements are often ordered in terms of their size. Dispersed, individual households are at the base of the rural settlement hierarchy. At the next level are hamlets (Figure 5). A hamlet is a small collection of farms and houses, which generally lacks all but the most basic services and facilities. The trade generated by the population, which is often less than 100 people, will only support **low order services** such as a general store, a sub post office or a pub. By contrast, a village has a much larger population. It can support a wider range of services, including school, church or chapel, community centre and a small range of shops (Figure 6). Higher up the hierarchy are towns and cities offering many more services and different types of services. As Figure 5 shows, there are more settlements lower down the hierarchy – the higher up you go, the fewer the number of each type of settlement. Thus, for example, there are far fewer cities in a country than there are villages.

Rural settlements offer certain functions and services. Only basic or **low order** functions are found

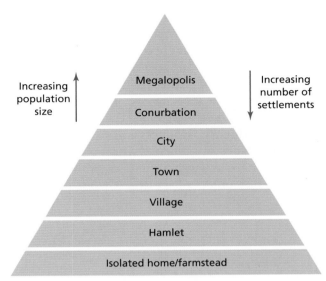

Figure 5 A hierarchy of settlement

in the smaller hamlet whereas the same functions and services are found in larger settlements (villages and market towns) together with more specialised ones – **high order** functions. Market towns draw custom from the surrounding villages and hamlets as well as serving their own population.

The maximum distance that a person is prepared to travel to buy an item (good) is known as the **range of a good**. Low order goods have a small range whereas high order goods have a large range.

Hamlet	Village	Small market town
General store	General store	General store
Post office	Post office	Post office
Pub	Pub	Pub
	Butcher	Butcher
	Garage	Garage
	Grocer	Grocer
	Hardwear	Hardwear
	Primary school	Primary school, baker, bike shop, chemist, confectionery, dry cleaner, electrical, TV/radio, furniture, hairdresser, laundrette, local government offices, off-licence, photo shop, restaurant, shoe shop, solicitor, supermarket, undertaker

Figure 6 A simple rural hierarchy

The number of people needed to support a good or service is known as the **threshold** population. Low order goods may only need a small number of people (1000 say) to support a small shop, whereas a large department store might require 50,000 people in order for it to survive and make a profit.

The area that a settlement serves is known as the **sphere of influence**. Hamlets and villages generally have low spheres of influence whereas larger towns and cities have a large sphere of influence. The definition of hamlet, village and town is not always very clearcut and these terms represent features that are part of a sliding scale (**continuum**) rather than separate categories.

In general, as population size in settlements increases the number and range of services increases (Figure 7). However, there are exceptions. Some small settlements, notably those with a tourist-related function, may be small in size but have many services. In contrast, some dormitory (commuter) settlements may be quite large but offer few functions or services other than a residential one. In these settlements, people live (reside) in the village but work and shop elsewhere.

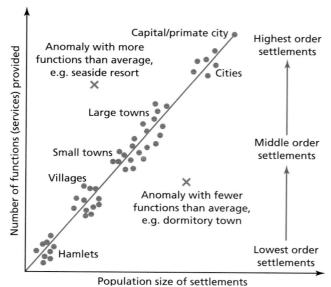

Figure 7 The relationship between population size and number of services

■ Factors affecting the size, growth and function of settlements

A number of factors affect settlement size, growth and function. In extreme environments settlements are generally small. This is because the environment is too harsh to provide much food. Areas that are too hot or cold, wet or dry usually have small, isolated settlements. In contrast, in areas where food production is favoured, settlements have managed to grow. If there is more food produced than the farmers need, then non-farming services can be supported. In the early days these included builders, craftsmen, teachers, traders, administrators and so on. Thus, settlements in the more favoured areas had greater potential for growth, and a greater range of services and functions.

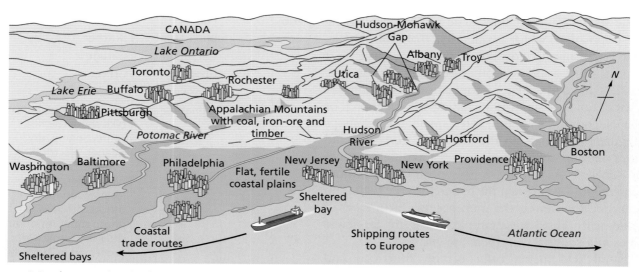

Figure 8 Settlement sites in the north-east of the USA

Figure 9 Water continues to play an important role in modern New York

Figure 10 The Seine was a vital factor in enabling the growth of Paris into a world city

Some environments naturally favoured growth. In the north-east of the USA, settlements on the lowland coastal plain were able to farm and trade (Figure 8). Those that had links inland as well, such as New York, were doubly favoured (Figure 9).

Trade and communications have always been important. Cairo grew as a result of being located at the meeting point of the African, Asian and European trade routes. It also benefited from having a royal family, being a seat of government, and having a university and all kinds of linked industries such as food and drink, and textiles. Similarly, Paris grew because of its excellent location on the Seine. Not only could the river be crossed at this point, it could also be used for trade (Figure 10).

Other centres had good raw materials. In South Africa, the gold deposits near Johannesburg, and

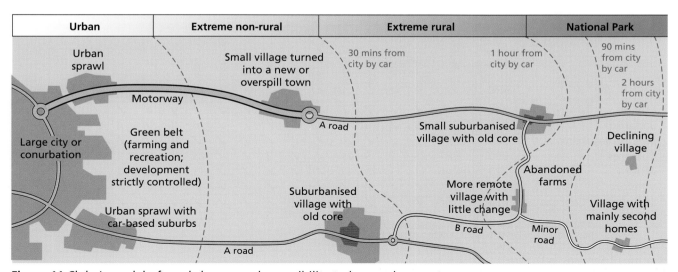

Urban	Extreme non-rural	Extreme rural	National Park

Urban sprawl

Small village turned into a new or overspill town

30 mins from city by car

1 hour from city by car

90 mins from city by car

Motorway

2 hours from city by car

A road

Green belt (farming and recreation; development strictly controlled)

Large city or conurbation

Small suburbanised village with old core

Declining village

Abandoned farms

More remote village with little change

Urban sprawl with car-based suburbs

Suburbanised village with old core

Village with mainly second homes

B road

Minor road

A road

Figure 11 Cloke's model of rural change and accessibility to large urban centres

the diamonds at Kimberley and Bloemfontein, enabled these settlements to grow as important mining and industrial areas.

Functions change over time. Many settlements that were formerly fishing villages have become important tourists resorts. The Spanish costas are a good example. Many Caribbean settlements, such as Soufrière in St Lucia, have evolved into important tourist destinations. In MEDCs many rural settlements have now become dormitory settlements – this is related to good accessibility to nearby urban centres (Figure 11, page 31). Increasingly, many rural settlements in MEDCs are also becoming centres of industry, as new science parks locate in areas such as Silicon Valley in California, formerly an agricultural region. In South Korea, the rural population has declined from 28 per cent in 1994 to less than 10 per cent in 2008 as

Year	Population	Year	Population
1801	130,000	1921	108,000
1821	135,000	1941	94,000
1841	142,000	1961	82,000
1861	138,000	1981	74,000
1881	144,000	2001	76,000
1901	130,000		

Figure 12 Population change at Lozère

the country has industrialised and urbanised over the last fifty years or so.

Other centres have become important due to political factors. New capital cities such as Brasilia, Canberra and Ottawa have developed important administrative roles. Other planned cities, such as Putrajaya in Malaysia and Icheong in South Korea, have become important centres for high-tech industry.

Activities

Figure 12 shows population data for the Lozère District, a remote upland area in southern France.

1 **a)** Draw a line graph to show the change in population between 1801 and 2001.
   b) Describe the changes in population in the graph you have drawn.
   c) Suggest reasons to explain the changes in population between:
      i) 1801 and 1881
      ii) 1881 and 1981
      iii) 1981 and 2001.

2 Figure 13 shows data for services in seven settlements in Lozère.
   a) Choose a suitable method to plot population size against number of services.
   b) Describe the relationship between population size and number of services for the region.
   c) Identify one exception to the pattern and suggest how, and why, it does not fit the pattern.
   d) Suggest a hierarchy of settlements based on the information provided.

Settlement	Altitude in metres	Population	Railway	Doctor	Chemist	Dentist	Restaurant	Hotel	Post office	Shops	Mobile shop	Cinema	Swimming pool	Swimming (river, lake)	Tennis	Fishing	Canoeing	Horse riding	Skiing
Mende	750	11,000	✓	✓	✓	✓	✓	✓	✓	✓	✓	✓	✓	✓	✓	✓	✓	✓	25
Badaroux	800	939	0.5	✓	6	6	✓	✓	✓	✓	✓	6	6	✓	4	✓	6	✓	12
Bagnois-les-Bains	913	240	6	✓	✓	20	✓	✓	✓	✓	✓	20	✓	✓	✓	✓	16	14
Cubières	900	281	25	9	9	25	✓	✓	9	✓	✓	25	25	20	9	✓	25	25	9
Altier	725	246	11	11	11	11	✓	✓	✓	✓	✓	11	11	11	11	✓	11	16	25
Villefort	605	791	✓	✓	✓	✓	✓	✓	✓	✓	✓	✓	✓	✓	✓	✓	✓	✓	✓
St André de Capcèze	450	104	✓	✓	✓	✓	✓	✓	✓	✓	✓	✓	✓	✓	✓	✓	✓	✓	14

Key Services available to tourists and residents in settlement: 3
Numbers show distance in km to nearest service, i.e. 25 = 25 km distant

Figure 13 Services in Lozère

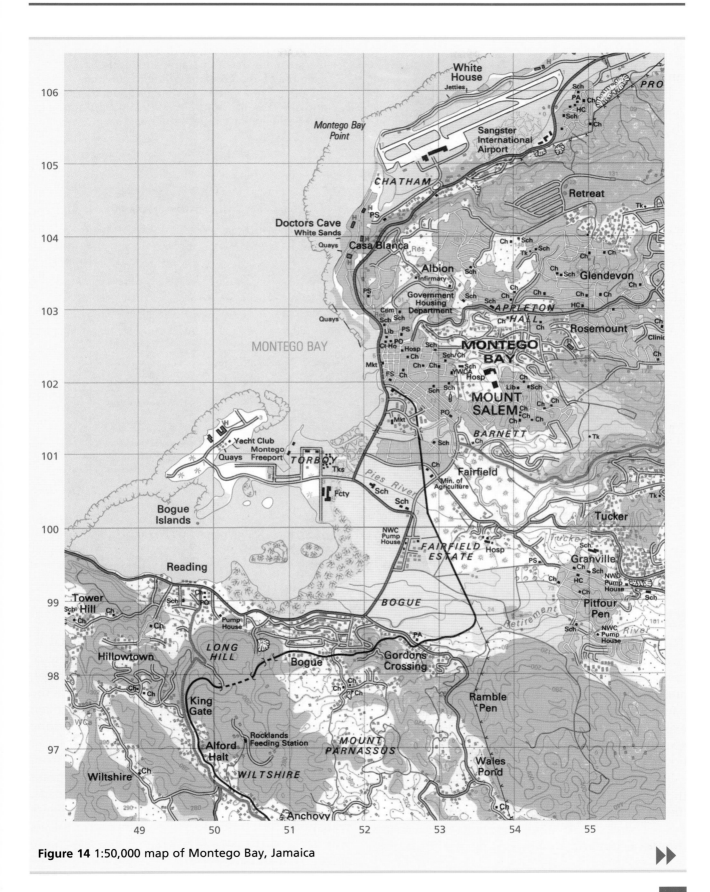

Figure 14 1:50,000 map of Montego Bay, Jamaica

Activities

Study Figure 14.

1. What is the grid square reference of a) the hospital in Montego Bay and b) the factory on Torboy (Bogue Islands)?

2. What is the square reference for a) the golf course north-east of Montego Bay and b) the Fairfield Estate?

3. What is the length of the longest runway at Sangster International Airport?

4. **a)** How far is it 'as the crow flies' (in a straight line) from the hotel (H) on Bogue Islands to the main buildings at Sangster International Airport?
   b) How far is it, by road, from the hotel (H) on Bogue Islands to the main buildings at Sangster International Airport?

5. In which direction is Gordons Crossing from the settlement of Montego Bay?

6. Describe the site of Montego Bay. Suggest why the area grew into an important tourist destination.

7. What types of settlement are found at Pitfour Pen (5598) and Wales Pond (5296)?

8. Suggest reasons for the lack of settlements in squares 5497 and 5199.

9. Suggest reasons for the growth of settlements at Bogue (5198) and Granville (5599).

10. Find an example of **a)** dispersed settlement and **b)** nucleated settlement on the map. Suggest why each type of settlement has that pattern in the area that it is found.

11. Using the map extract, work out a settlement hierarchy for the area. Name and locate an example of **a)** an area of isolated, individual buildings, **b)** a village, **c)** a minor town, **d)** a town and **e)** a large town. Use the key (Figure 15) to help you decide what type of settlement each one is.

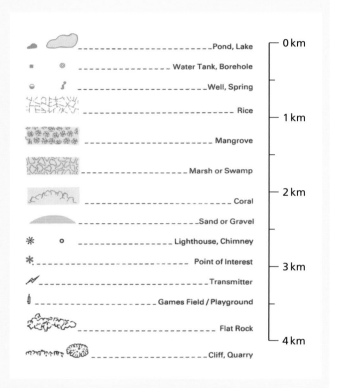

Figure 15 Key to 1:50,000 map of Montego Bay

Urban land use

The growth of cities in the nineteenth and early twentieth centuries produced a form of city that was easily recognisable. It included a central commercial area, a surrounding industrial zone with densely packed housing, and outer zones of suburban expansion and development. Geographers have spent a lot of time modelling these cities to explain how they 'work'.

Every model is a simplification. No city will 'fit' these models perfectly, but there are parts of every model that can be applied to most cities in the developed world. All models are useful in as much as they focus our attention on one or two key factors.

Bid-rent theory

The concept of **bid rent** (Figure 16a) is vital to models of urban land use. Bid rent is the value of land for different purposes, such as commercial, manufacturing and residential purposes. Land at the centre of a city is the most expensive for two main reasons: it is the most **accessible** land to **public transport**, and there is only a small amount available. Land prices generally decrease away from the central area, although there are secondary peaks at the intersections of main roads and ring roads. Change in levels of accessibility, due to **private transport** as opposed to public transport, explains why areas on the edge of town are often now more accessible than inner areas.

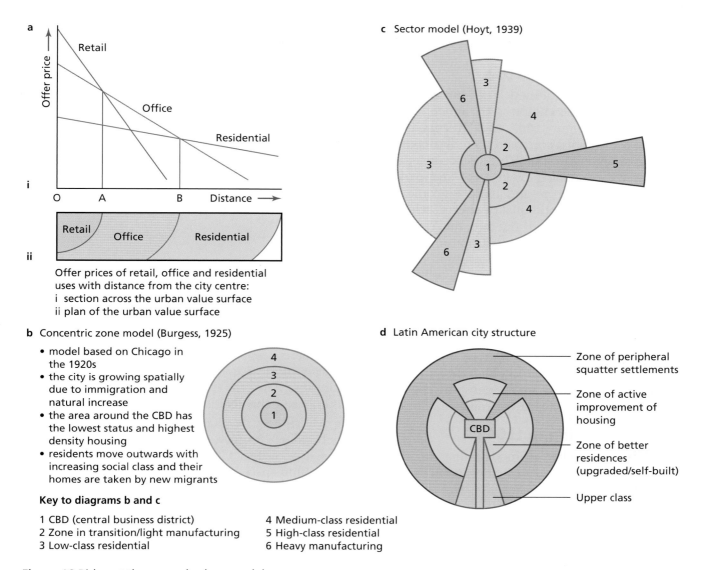

a

Retail

Office

Residential

Offer price

i

O　　A　　　　B　　Distance →

ii

Retail　Office　Residential

Offer prices of retail, office and residential
uses with distance from the city centre:
i section across the urban value surface
ii plan of the urban value surface

b Concentric zone model (Burgess, 1925)

- model based on Chicago in
  the 1920s
- the city is growing spatially
  due to immigration and
  natural increase
- the area around the CBD has
  the lowest status and highest
  density housing
- residents move outwards with
  increasing social class and their
  homes are taken by new migrants

Key to diagrams b and c

1 CBD (central business district)　　4 Medium-class residential
2 Zone in transition/light manufacturing　5 High-class residential
3 Low-class residential　　　　　　　6 Heavy manufacturing

c Sector model (Hoyt, 1939)

d Latin American city structure

CBD

Zone of peripheral
squatter settlements

Zone of active
improvement of
housing

Zone of better
residences
(upgraded/self-built)

Upper class

Figure 16 Bid-rent theory and urban models

Burgess's concentric model (1925)

This is the basic model (Figure 16b). Burgess
assumed that new migrants to a city moved into
inner city areas where housing was cheapest and it
was closest to the sources of employment. Over
time residents move out of the inner city area as
they become wealthier. In his model, housing
quality and social class increase with distance from
the city centre. Land in the centre is dominated by
commerce as it is best able to afford the high land
prices, and requires highly accessible sites. In the
early twentieth century, public transport made the
central city the most accessible part of town.
Beyond the central business district (CBD) is a

manufacturing zone that also includes high density,
low quality housing to accommodate the workers.
As the city grows and the CBD expands, the
concentric rings of land use are pushed further out.
The area of immediate change adjacent to the
expanding CBD is known as the zone in transition
(usually from residential to commercial).

Hoyt's sector model (1939)

Homer Hoyt emphasised the importance of
transport routes and the incompatibility of certain
land uses (Figure 16c). Sectors develop along
important routeways, while certain land uses, such as
high-class residential and manufacturing industry,

deter each other and are separated by buffer zones or physical features.

Urban land use in LEDCs

There are a number of models of Third World cities. One of the most common is the model of a Latin American city (Figure 16d). The CBD has developed around the colonial core, and there is a commercial avenue extending from it. This has become the spine of a sector containing open areas and parks, and homes for the upper- and middle-income classes. These areas have good-quality streets, schools and public services. Further out are the more recent suburbs, with more haphazard housing and fewer services. More recent squatter housing is found at the edge of the city. Older and more established squatter housing is found along some sectors that extend in towards the city centre. Conditions in these areas near the city centre are better than in the more recent areas at the edge. In addition, those living in the central areas are closer to centres of employment and are more likely to find work. Industrial areas are found scattered along major transport routes, with the latest developments at the edge.

Land use zoning in LEDCs

There are a number of models that describe and explain the development of cities in LEDCs. There are several key points:

- The rich generally live close to the city centre whereas the very poor are more likely to be found on the periphery.

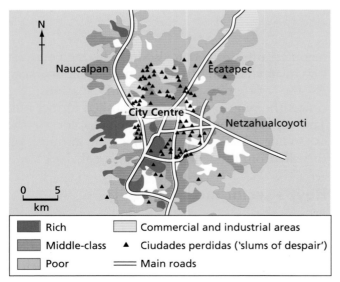

Figure 17 Urban land use in Mexico City

- Higher-quality land is occupied by the wealthy.
- Segregation by wealth, race and ethnicity is evident.
- Manufacturing is scattered throughout the city.

Activities

1 Describe the pattern of land use in Mexico City as shown in Figure 17.

2 To what extent does the pattern of land use in Figure 17 conform to:

   a) the model of land use in Figure 16d
   b) any of the characteristics described in Figure 16a–c?

■ Internal structure of towns and cities

The central business district

The central business district (CBD) is the commercial and economic core of the city, the area that is most accessible to public transport and the location with the highest land values. It has a number of characteristic features:

- Multistorey development – high land values force buildings to grow upwards; hence the total floor space of the CBD is much greater than the ground space.
- Concentration of retailing – high levels of accessibility attract shops with high range and threshold characteristics, such as department stores in the most central areas, while specialist shops are found in less accessible areas.
- Concentration of public transport – there is a convergence of bus routes on the CBD.
- Concentration of offices – centrality favours office development.
- Vertical zoning – shops occupy ground floors because of accessibility while offices occupy upper floors.
- Functional grouping – similar shops and similar functions tend to locate together (increasing their thresholds).
- Low residential population – high bid rents can only be met by luxury apartments.
- Highest pedestrian flows – this is due to the attractions of a variety of commercial outlets and service facilities.
- Traffic restrictions are greatest in the CBD – pedestrianisation has reduced access for cars since the 1960s.

- The CBD changes over time – there is an assimilation zone (the direction in which the CBD is expanding) and there is a discard zone (the direction from which it is moving away).

There are, however, many problems in the CBD, such as a lack of space, the high cost of land, congestion, pollution, a lack of sites, planning restrictions, and strict government control.

The core–frame concept

This concept suggests that the CBD (core) is surrounded by a zone in which there are specialised services, such as medical, administrative and educational, wholesale and warehousing, transport and light manufacturing (Figure 18). The CBD and the frame are closely connected and the CBD core may advance into the frame, just as the frame may

Core and frame elements of the CBD

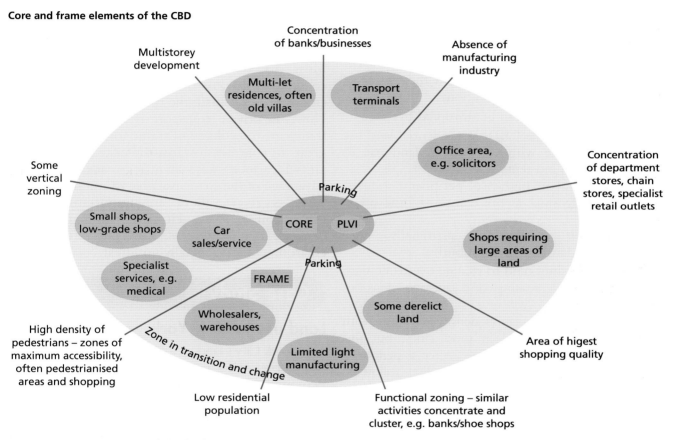

PLVI = peak land value intersection: the highest rated, busiest, most accessible part of a CBD

Factors influencing CBD decline

Figure 18 The core–frame model

advance into the core as parts of the CBD become run down. This is similar to the concept of zones of assimilation and discard.

Residential zones

In most MEDCs, as a general rule, residential densities decrease with distance from the CBD. This is due to a number of reasons:

- historically, more central areas developed first and supported high population densities
- greater availability of land with increased distance from the CBD
- improvements in transport and technology allow people to live further away from their place of work in lower-density conditions.

However, this pattern can be disrupted by:

- low densities in the CBD, as residential land use cannot compete with commercial land use to meet the high bid rents
- the location of high-rise peripheral estates, increasing densities at the margins of the urban area
- 'green belt' restrictions artificially raise population densities in the suburbs.

Population densities tend to change over time with peak densities decreasing and average densities increasing.

The pattern of population density declining with distance can be observed in most cities, but this pattern also changes over time. After a period of expansion, city centres start to decline following suburbanisation. This is sometimes followed by a repopulation of the inner city if the centre is redeveloped.

Industrial areas

There are a number of industrial zones in most MEDC cities. These include:

- traditional inner city areas close to railways and/or canals
- areas that require access to water, for example dock-related industries such as imports and exports
- areas where there is good access and good availability of land
- edge-of-town/greenfield suburban sites close to airports.

Open spaces

In general the amount of open space increases towards the edge of town. This is because the value of land is less towards the edge, and there is more land available. Nevertheless, there are important areas of open space in many urban areas. Central Park in New York is a good example. In the centre, the areas of open land tend to be small. Many of the open spaces are related to areas that are next to rivers or formerly belonged to wealthy land-owners.

Transport routes

Most city centres are characterised by narrow, congested roads. As the roads were built when the cities were still small, the roads were quite narrow. Now, as private transport is the main form of transport, the volume of traffic for the roads is too great. In contrast, towards the edge of town there are wider motorways and ring-roads. These take advantage of the space available. Natural routeways such as river valleys are important for the orientation of roads. However, given that many cities are in lowland areas, constraints of the natural environment are generally not great.

Case Study

Land use in New York

New York City's land area covers about 825 km^2. The distribution of commercial land use is dominated by two areas, Midtown Manhattan and Downtown (Lower) Manhattan (Figures 19a–c, pages 39–41). Lower Manhattan is the centre for finance and banking, containing Wall Street and the Stock Exchange. In contrast, midtown Manhattan has the main shops (Fifth Avenue), theatres (Broadway, hotels and landmark buildings such as the Rockefeller Center and the Empire State Building. Commercial areas occupy less than 4 per cent of the city's land, but they use space intensively. Most of the city's 3.6 million jobs are in commercial areas, ranging from the office towers of Manhattan and the regional business districts of downtown Brooklyn, to the local shopping corridors throughout the city.

Industrial uses, the warehouses and factories occupy 4 per cent of the city's total lot area. They are

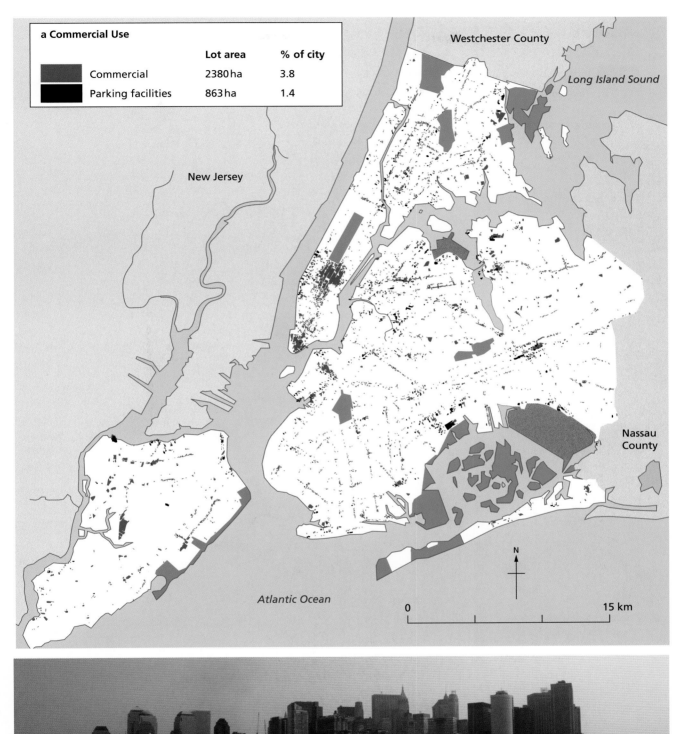

Figure 19a Commercial land use in New York

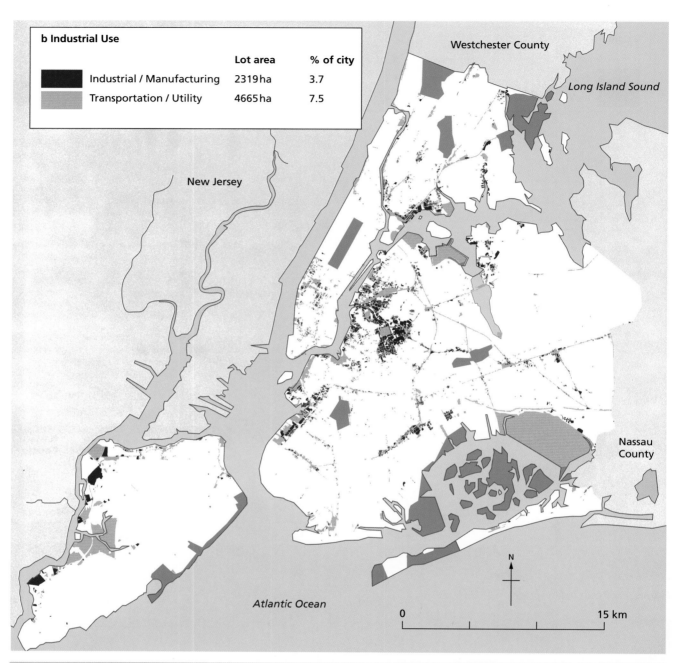

b Industrial Use

	Lot area	% of city
Industrial / Manufacturing	2319 ha	3.7
Transportation / Utility	4665 ha	7.5

Westchester County

Long Island Sound

New Jersey

Nassau County

Atlantic Ocean

N

0   15 km

Figure 19b Industrial land use in New York

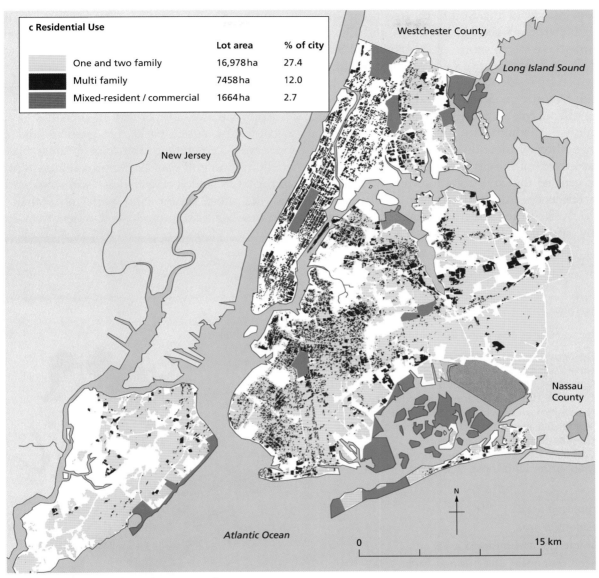

Figure 19c Residential land use in New York

	1	2	3	4	5	6	7	8	9	10	11	12
Bronx	18.1	15.5	2.7	4.3	3.8	2.8	11.6	31.1	2.0	4.3	3.8	100
Brooklyn	22.7	16.0	3.4	3.1	4.9	4.2	6.0	33.9	1.8	3.1	0.9	100
Manhattan	1.3	23.9	12.2	10.2	2.4	6.6	11.7	25.1	1.7	3.0	1.9	100
Queens	36.2	10.6	1.5	3.2	3.7	11.8	4.5	19.7	1.3	5.2	2.3	100
Staten Island	33.6	3.1	0.5	3.4	2.9	7.8	9.6	20.7	0.5	17.5	0.5	100
New York City	27.4	12.0	2.7	3.8	3.7	7.5	7.3	25.4	1.4	6.9	1.8	100

Key

1 Low-density residential areas
2 High-density residential
3 High-density apartments/commercial
4 Commercial/office

5 Industrial/manufacturing
6 Transport/utility
7 Public facilities and institutions
8 Open space

9 Parking facilities
10 Vacant land
11 Miscellaneous
12 Total

Figure 20 Land use in New York City by borough (percentage figures)

found primarily in the South Bronx, along either side of Newtown Creek in Brooklyn and Queens, and along the western shores of Brooklyn and Staten Island. River-front locations are very important.

Low-density residences, the largest use of city land, are found mostly in Staten Island, eastern Queens, southern Brooklyn, and north-west and eastern Bronx. In contrast, medium- to high-density residential buildings (three or more dwelling units) contain more than two-thirds of the city's housing units but occupy 12 per cent of the city's total lot area. The highest-density residences are found mainly in Manhattan, and four- to twelve-storey apartment houses are common in many parts of the Bronx, Brooklyn and Queens.

Public facilities and institutions – including schools, hospitals and nursing homes, museums and performance centres, houses of worship, police and fire stations, courts and detention centres – are spread throughout the city and occupy 7 per cent of the city's land.

Approximately 25 per cent of the city's lot area is occupied by public parks, playgrounds and nature preserves, cemeteries, amusement areas, beaches, stadiums and golf courses. Approximately 8 per cent of the city's land is classified as vacant. Staten Island has the most vacant land with more than 2150 hectares, Manhattan the least with less than 200.

Activities

1 Refer to Figures 19a–c and 20 (pages 39–41).

a) Describe the distribution of commercial land in New York City.

b) Comment on the distribution of industrial land.

c) Compare the distribution of low-density residential areas with that of high-density residential areas. How do these compare with the distribution of apartments?

d) Compare and contrast the main land uses in Manhattan with those of Queens.

Urbanised area	
Residential area	6.9
Commercial and business area	0.9
Mixed residential and business area	0.9
Industrial area	1.4
Public facilities	0.1
Transport	10.2
Urban infrastructure facilities	1.3
Construction sites and wasteland	2.4
Total	**58.1**
Forest and open space	
Grassland	2.9
Farmland	4.9
Rivers and wetland	5.8
Forest	26.2
Unsurveyable	2.1
Total	**41.9**

Figure 21 Urban land use in Seoul (%)

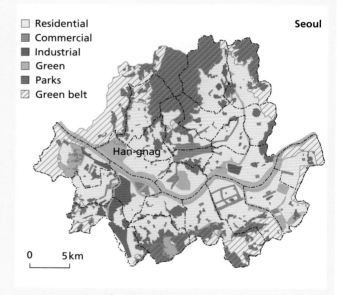

Legend:
- ☐ Residential
- ▨ Commercial
- ■ Industrial
- ▨ Green
- ▨ Parks
- ▨ Green belt

Seoul

Han-gnag

0   5 km

Figure 22 Land use in Seoul

2 a) Using the data in Figure 21, draw a pie chart to show land uses in Seoul.

b) Comment on your diagram.

c) Describe the distribution of industrial land use as shown in Figure 22.

d) Describe, and suggest reasons for, the distribution of open space, as shown in Figure 22.

Problems associated with urban growth

A number of problems are associated with the growth of urban areas. These include:

- congestion in the CBD
- very high land prices in the city centre
- overcrowding
- housing shortages
- traffic congestion
- unemployment
- racial conflict
- urban decay and dereliction
- deprivation
- pollution of air and water.

These problems are found in most large cities in MEDCs and LEDCs. LEDCs have the added problems of shanty housing and squatter settlements.

Case Study

Urban problems in New York

New York's population is declining and changing:

- The total population has fallen by over 10 per cent in the last decade.
- The white population has fallen from 87 per cent to 65 per cent since 1950.
- The middle-class population has fallen by 2 million in the past 20 years.
- The number of elderly residents has risen by 21 per cent since 1950.
- New York has the largest black population in the country, 9.1 per cent.
- Population of the South Bronx has declined by nearly 50 per cent.
- Almost 1 million New Yorkers are on welfare.

The inner areas are declining both in terms of population and employment. Much of the housing and industry is dated (Figure 23). By contrast, the suburban periphery has expanded in terms of population and employment. Only 20 per cent of the workers in Westchester County (a New York suburb and one of the wealthiest counties in the USA) now commute into the city for their work.

Within New York there is much poverty. Up to 25 per cent of its citizens now live in poverty. A good example of an inner city area is the South

Figure 23 Old tenement housing in New York

Bronx. There, the average income is 40 per cent that of the country as a whole, and one-third of residents are on welfare support. New York's problems arise from:

- changes in the composition of the labour force
- the high living standards which the US economy guarantees for most citizens
- the social strains set up by the continuing existence of an undereducated, unskilled, underpaid and underprivileged minority
- the massive outward movement of the middle class, spurred by good roads and increased living standards
- the counter-movement of lower income families into the inner city.

Case Study

Problems in Seoul
Housing shortage

Seoul's population has grown from 2.5 million in 1960 to over 10 million in 1990 and over 20 million in the Greater Seoul region by 2005. In-migration and the trend towards nuclear families (two generations) rather than the extended family (three generations in the one house) have created a major housing shortage, despite massive building programmes (Figure 24). Less than 45 per cent of the land around Seoul is available for urban development due to steep terrain and mountains. The type of housing is changing too. The typical one-storey, one-family house with inner courtyards is being replaced by high-rise apartment blocks. Such flats have increased from 4 per cent of housing in 1970 to 35 per cent in 1990 and 50 per cent in 2005. Until recently most of the housing was to the north of the river, but a number of satellite towns have been built to the south of the river. This has evened out population density, which is, on average, over 16,000 people per km² (Figure 25).

Traffic congestion

Like New York, Seoul experiences massive traffic congestion. In 1975 Korea manufactured fewer than 20,000 cars, but by 1994 there were over 2 million cars registered in the Seoul area. Despite improvements to the motorway network, the increase in the population of Seoul and the number of cars in the area means that congestion has increased. In addition, many of the roads in central Seoul are relatively small and unable to handle the large volumes of traffic.

Figure 24 New buildings in Seoul

	People/km²
Seoul	16,364
Tokyo	13,092
Beijing	4,810
Singapore	4,773
London	4,671
Paris	8,084
New York	9,721
Los Angeles	3,037

Figure 25 Population densities in selected cities

Figure 26 The Cheong Gye Cheong River in central Seoul, c1950

Pollution

As Seoul has grown, the amount of air and water pollution has increased. A good example was the Cheong Gye Cheong River in central Seoul (Figure 26). It had become heavily polluted with lead, chromium and manganese and was a health risk. Restoration of the river has been a central part of the regeneration of central Seoul. Previously up to 87 per cent of the city's sewage flowed untreated into the Hangang River. Now Seoul has the capacity to treat up to 3 million tonnes of sewage each day.

Activities

1 Suggest contrasting reasons why there is poverty in New York.

2 Why is air pollution a problem in large cities?

3 Describe the conditions around the Cheong Gye Cheong River as shown in Figure 26.

Urban problems in LEDCs

Major area/region	Total population (millions)	Urban population (millions)	Urban population (% of total population)	Estimated slum population (thousands)	Estimated slum population (% of urban population)
World	6134	2923	47.7	923,986	31.6
Developed regions	1194	902	75.5	54,068	6.0
Europe	726	534	73.6	33,062	6.2
Other	467	367	78.6	21,006	5.7
Developing regions	4940	2022	40.9	869,918	43.0
Northern Africa	146	76	52.0	21,355	28.2
Sub-Saharan Africa	667	231	34.6	166,208	71.9
Latin America and the Caribbean	527	399	75.8	127,567	31.9
Eastern Asia	1364	533	39.1	193,824	36.4
South-central Asia	1507	452	30.0	262,354	58.8
South-eastern Asia	530	203	38.3	56,781	28.0
Western Asia	192	125	64.9	41,331	33.1
Oceania	8	2	26.7	499	24.1
Least developed countries	685	179	26.2	140,114	78.2
Landlocked developing countries	275	84	30.4	47,303	56.5
Small island developing states	52	30	57.9	7,321	24.4

Figure 27 Slum population by region, 2001

Case Study

Urban problems in Rio de Janeiro

Rio de Janeiro is a city of contrasts with a huge gap between rich and poor.

Rio's urban problems and their causes

Rio de Janeiro has many slums or *comunidade*. The natural increase in population is much higher in the most recent favelas, which are largely in the outer suburbs and on the rural–urban fringe. As spontaneous settlements are forced to develop on available land, most of the sites have been used in the central and inner urban areas. Many favelas were moved to outer suburban areas. In inner urban areas, newly established favelas were frequently forced to develop on steep hillsides, where landslides are a threat. About 17 per cent of Rio's population are favela-dwellers. They occupy just 6.3 per cent of Rio's land (Figure 28, page 46).

There are four main types of slums in Rio:

- squatter settlements or favelas – dense invasions of land with self-built housing on land lacking in infrastructure
- illegal subdivisions of land and/or housing
- invasions – irregular occupations of land still in the process of becoming fully established; these are found in 'risky' areas such as under viaducts or electricity lines, on the edge of railways or in public streets and squares
- corticos – old decaying housing that has been rented out without any legal basis; these are mainly located in the central areas and the port area.

Population growth is very rapid in some slums. For example, Rio das Pedras, located in a flooded swamp area, grew to 18,000 within its first two years.

Recently there has been decentralisation of activities from the CBD towards the edge of town. This has been mirrored by the movement of high-income classes to the coastal expansion areas in the east of the city.

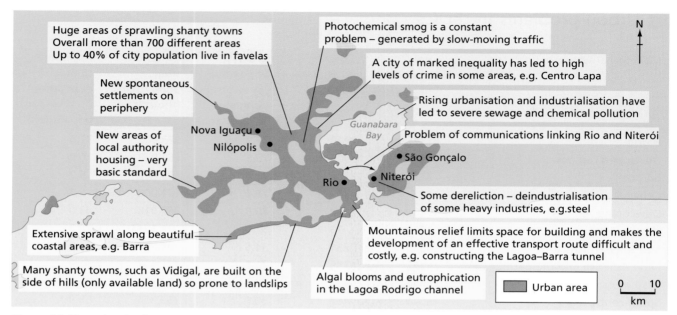

Figure 28 Slums in Rio de Janeiro

The inequality in wealth in Rio is staggering – the richest 1 per cent of the population earns 12 per cent of the income and the poorest 50 per cent earns just 13 per cent of the city's income. The southern zone of the city is the richest area.

Rio's period of extremely rapid growth (1960–80) has slowed, yet growth is still at the rate of over a million a decade. Public services such as education and health have become inadequate as a result of rapid urban growth. The steep mountains that surround the narrow, flat coastal strips of land have affected the physical growth of Rio. The mountainous relief limits the space for building and makes the development of an effective transport network more difficult. The few existing transport routes have to be used by everyone, which leads to traffic congestion. The mountains surrounding the city trap photochemical smog created by exhaust fumes, resulting in poor air quality.

Raw sewage has been draining straight into the bay, with population growth and industrial growth in Rio intensifying the problems. Today there is little marine life left in the bay, commercial fishing has decreased by 90 per cent in the last 20 years, and swimming from beaches in the interior of the bay is not advisable. The Lagoa Rodrigo de Freitas is the lake inland from Ipanema and Leblon. The release of raw sewage leads to algae blooms and eutrophication which resulted in the death of 132 tonnes of fish in the Lagoa in February 2000.

Activities

1 Refer to Figure 27 (page 45).

   a) Choose an appropriate method to show the global distribution of populations living in slums.
   b) Where is the frequency of slum dwellers highest? Where is it lowest?
   c) Comment on the results you have drawn.

2 Suggest reasons why there is a housing shortage in Rio de Janeiro.

3 Describe one environmental problem that is the result of rapid urban growth.

4 Explain why natural hazards may have a major impacts on areas of slum housing.

Solutions to urban problems

The housing crisis in LEDCs

Provision of enough quality housing is also a major problem in LEDCs. There are at least four aspects to the management of housing stock:

Figure 29 Poor-quality housing in Oisins, Barbados

- quality of housing (Figure 29) – with proper water, sanitation, electricity, and space
- quantity of housing – having enough units to meet demand
- availability and affordability of housing
- housing tenure (ownership or rental).

There are a variety of possible solutions to the housing problem:

- government support for low-income self-built housing
- subsidies for home building
- flexible loans to help shanty-town dwellers
- slum upgrading in central areas
- improved private and public rental housing
- support for the informal sector/small businesses operating at home
- site and service schemes
- encouragement of community schemes
- construction of health and educational services.

Housing the urban poor

LEDC governments would not be able to solve their housing problems even if they were to try. The best that they could be expected to do in an environment of general poverty is to improve living conditions. They should try to:

- reduce numbers of people living at densities of more than 1.5 persons in each room
- increase access to electricity and potable water
- improve sanitary facilities
- prevent families moving into areas that are physically unsafe

- encourage households to improve the quality of their accommodation.

A sensible approach is to destroy slums as seldom as possible, on the grounds that every displaced family needs to be rehoused and removing families is often disastrous. Governments should also avoid building formal housing for the very poor. Sensible governments will attempt to upgrade inadequate accommodation by providing it with infrastructure and services of an appropriate standard.

There are no easy solutions to LEDC housing problems because poor housing is merely one manifestation of generalised poverty. Decent shelter can never be provided while there is widespread poverty.

Urban agriculture

The phrase 'urban agriculture' initially sounds like a contradiction in terms; however, the phenomenon has grown in significance in the cities of LEDCs over the past 20 years. Evidence suggests that in some cities urban agriculture may already occupy up to 35 per cent of the land area, may employ up to 36 per cent of the population, and may supply up to 50 per cent of urban fresh vegetable needs.

Food produced locally in urban areas may have several added benefits. First, it employs a proportion of the city's population. Secondly it can be cheaper for urban dwellers to grow at least some of their own food – in Addis Ababa dwellers can save between 10 and 20 per cent of their income through urban cultivation. Third, it diversifies the sources of food, resulting in a more secure supply.

Advantages	Concerns
Vital or useful supplement to food procurement strategies	Conflict over water supply, particularly in arid or semi-arid areas
Various environmental benefits	Health concerns, particularly from use of contaminated wastes
Employment creation for the jobless	Conflicting urban land issues
Provides a survival strategy for low-income urban residents	Focus on the urban cultivation activities rather than in relation to broader urban management issues
Urban agriculture makes use of urban wastes	Urban agriculture can benefit only the wealthier city dwellers in some cases

Figure 30 Urban agriculture – the pros and cons

Figure 31 Gongju-Yongi, the proposed new capital of South Korea – planning board and in 2007

New cities

For rich countries, there are more options. At one end of the scale are new towns and cities, such as Brasilia, Canberra and, in Korea, Gongju-Yongi. Gongju-Yongi is a £26 billion scheme to reduce the importance of Seoul as Korea's capital by 2020 (Figure 31). The relocation is necessary to ease chronic overcrowding in Seoul, redistribute the state's wealth, and reduce the danger of a military attack from North Korea. Previous developments have concentrated huge amounts of money, power and up to half of Korea's population in Seoul. Construction on the new city began in 2007. Another impressive scheme is the Malaysian new town of Putrajaya.

■ Managing urban problems

Managing transport in cities

MEDCs	LEDCs
■ Increased number of motor vehicles	■ Private car ownership is lower
■ Increased dependence on cars as public transport declines	■ Less dependence on the car, but growing
■ Major concentration of economic activities in CBDs	■ Many cars are poorly maintained and are high polluters
■ Inadequate provision of roads and parking	■ Growing centralisation and development of CBDs increases traffic in urban areas
■ Frequent roadworks	■ Heavy reliance on affordable public transport
■ Roads overwhelmed by sheer volume of traffic	■ Journeys are shorter but getting longer
■ Urban sprawl results in low-density built-up areas, and increasingly long journeys to work	■ Rapid growth has led to enormous urban sprawl and longer journeys
■ Development of out-of-town retail and employment leads to cross-city commuting	■ Out-of-town developments are beginning as economic development occurs, e.g. Bogota (Colombia)

Figure 32 Traffic problems in MEDC and LEDC cities

'Carrots'	'Sticks'
■ Park-and-ride schemes – parking at the terminal for a major bus or train route, e.g. Oxford (UK), Brisbane (Australia)	■ High car parking charges in city centres, e.g. Copenhagen (Denmark), London, Oxford
■ Subsidised public transport systems, e.g. Oxford, Zurich (Switzerland), Brisbane	■ Restricted city-centre parking, e.g. Copenhagen (Denmark), Cambridge (UK)
■ Modern electronic bus systems with consumer information on frequency, e.g. Brisbane, Curitiba (Brazil); rapid transit systems, e.g. supertrams on dedicated tracks, e.g. Zurich; underground trains, e.g. Newcastle (UK), Cairo (Egypt)	■ Road tolls and road pricing (congestion charges), e.g. Durham (UK), Bergen (Norway) and central London, so that people have to pay to drive through congested areas of the city centre
■ Providing bus lanes to speed up buses, e.g. Oxford, London	

Figure 33 Attempts to manage the transport issue

Case Study

Sustainable development in Curitiba

Curitiba, a city in south-west Brazil, is an excellent model for sustainable urban development (Figure 34). It has experienced rapid population growth, from 300,000 in 1950 to over 2.1 million in 1990, but has managed to avoid all the problems normally associated with it. This success is largely due to innovative planning:

- Public transport is preferred over private cars.
- The environment is used rather than changed.
- Cheap, low-technology solutions are used rather than high-technology ones.
- Development occurs through the participation of citizens (bottom-up development) rather than top-down development (centralised planning).

Figure 34 The location of Curitiba in Brazil

Transport

Transport in Curitiba is well integrated. The road network and public transport system have structural axes – the direction of the roads/areas along which growth may take place. These allow the city to expand but keep shops, workplaces and homes closely linked. There are five main axes of the three parallel roadways:

- express routes
- direct routes
- local roads.

Curitiba's mass transport system is based on the bus. Interdistrict and feeder bus routes complement the express bus lanes along the structural axes.

Problems (1950s/60s)	Solutions (late 1960s onwards)
Many streams had been covered to form underground canals which restricted water flow	Natural drainage was preserved – these natural floodplains are used as parks
Houses and other buildings had been built too close to rivers	Certain low-lying areas are off-limits
New buildings were built on poorly drained land on the periphery of the city	Parks have been extensively planted with trees; existing buildings have been converted into new sports and leisure facilities
Increase in roads and concrete surfaces accelerated runoff	Bus routes and bicycle paths integrate the parks into the urban life of the city

Figure 35 Sustainable solutions to flooding

Everything is geared towards the speed of journey and convenience of passengers.

- A single fare allows transfer from express routes to interdistrict and local buses.
- Extra-wide doors allow passengers to crowd on quickly.
- Double- and triple-length buses allow for rush-hour loads.

The rationale for the bus system was economic as well as being designed for sustainability. A subway would have cost $70–80 million per km but the express busways were only $200,000 per km. The bus companies are paid by the kilometres of road they serve, not the number of passengers. This ensures that all areas of the city are covered.

Case Study

Rio de Janeiro
Housing

Areas of spontaneous housing in Rio used to be bulldozed without warning. However, the authorities were unable to offer enough alternative housing, with the result that the favelas grew again. The authorities have now allowed such areas to become permanent (Figure 36).

Figure 36 A favela in Rio de Janeiro

The **Favela Bairro Project** (Favela Neighbourhood Project) began in Rio in 1994. It aimed to recognise the favelas as neighbourhoods of the city in their own right and to provide the inhabitants with essential services. Approximately 120 medium-sized favelas (those with 500–2500 households) were chosen. The primary phase of the project addressed the built environment, aiming to provide:

- paved and formally named roads
- water supply pipes and sewerage/drainage systems
- crèches, leisure facilities and sports areas
- relocation for families who were living in high-risk areas, such as areas subject to frequent landslides
- channelled rivers to stop them changing course.

The second phase of the project aimed to bring the favela dwellers into mainstream society and keep them free of crime. This is being done by:

- generating employment, for example by creating co-operatives of dressmakers, cleaners, construction workers, and helping them to get established in the labour market
- improving education and providing relevant courses such as ICT
- giving residents access to credit so that they can buy construction materials and improve their homes.

The project has been used as a model of its type. The government is also helping people to become home-owners.

The mountainous relief of Rio means there is not a great amount of building space available. Development has consequently moved out (decentralised) to create 'edge towns' such as Barra da Tijuca. Barra is an example of decentralisation of the rich and upper classes.

Education

A number of developments have taken place to try and improve the quality of the education system. The *Amigos da Escola* (school friends) initiative encourages people from the community to volunteer their skills to improve opportunities offered by their local schools. The *Bolsa Escola* (school grants) scheme gives monthly financial incentives for low-income families to keep their children at school.

Rocinha is a central favela with a population of about 200,000 inhabitants. Over 90 per cent of the buildings in Rocinha are now constructed from brick and have electricity, running water and sewerage systems. Rocinha has its own newspapers and radio station. There are food and clothes shops, video rental shops, bars, a travel agent and even a McDonald's.

Many of these improvements and developments are the result of Rocinha's location close to wealthy areas such as São Conrado and Copacabana. Many of the residents work in these wealthy areas that surround Rocinha, and although monthly incomes are low, they are not as low as elsewhere in the city and in Brazil. These regular incomes have allowed improvements to be carried out by the residents themselves.

Transport

An efficient **bus service**, which covers all areas of the city, has been developed and the prices are within reach of most people, being US$0.50 for each journey made. Besides the organised bus services, vans operate along the most popular routes and charge the same price.

A **Metro system** has been built and is currently being expanded (Figure 37). A one way ticket costs US$0.66. Currently the Metro does not provide an alternative mode of transport for much of the population, as its coverage is not extensive enough.

The **Linha Vermelha** and the **Linha Amarela** are two major roads that have been built to try and ease traffic congestion. However there are indications that the number of private cars on the road has increased since these roads were built, which will reduce the long-term impact they have on traffic congestion.

Water pollution

Much work has already taken place in Rio to improve the sewerage systems. Improving the sewage systems will also help revive the Lagoa Rodrigo de Freitas – 4 km of new sewage pipes are currently being installed around the lake.

◼ Urbanisation and the environment

Managing environmental problems

Environmental problems that most cities have to deal with include:
- water quality
- dereliction

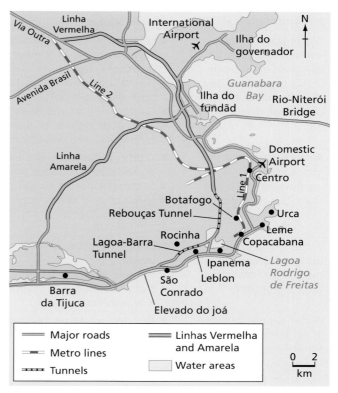

Figure 37 Transport systems in Rio de Janeiro

- air quality
- noise
- environmental health of the population.

There is a range of environmental problems in urban areas (Figure 38, page 52). These vary over time as economic development progresses. The greatest concentration of environmental problems occurs in cities experiencing rapid growth (Figure 39, page 52). This concentration of problems is referred to as the **Brown Agenda**. It has two main components:

- issues caused by limited availability of land, water and services
- problems such as toxic hazardous waste, pollution of water, air and soil, and industrial 'accidents' such as that at Bhopal in India in 1985.

Attempts to turn cities green can be expensive. Increasingly local governments are monitoring the environment to check for signs of environmental stress, and then applying some form of **pollution management**, **integrated management**, or **conservation order** to protect the environment.

Problems (and examples)	Causes	Possible solutions
Waste products and waste disposal – 25% of all urban dwellers in LEDCs have no adequate sanitation and no means of sewage disposal	■ Solids from paper, packaging and toxic waste increase as numbers and living standards rise ■ Liquid sewage and industrial waste both rise exponentially ■ Contamination and health hazards from poor systems of disposal, e.g. rat infestation and waterborne diseases	■ Improved public awareness – recycling etc., landfill sites, incineration plants ■ Development of effective sewerage systems and treatment plants including recycling of brown water for industrial use ■ Rubbish management
Air pollution – air in Mexico City is 'acceptable' on fewer than 20 days annually!	■ Traffic, factories, waste incinerators and power plants produce pollutants ■ Some specialist chemical pollution ■ Issues of acid deposition	■ Closure of old factories and importation of clean technology, e.g. filters ■ Use of cleaner fuels ■ Re-siting of industrial plants, e.g. oil refineries in areas downwind of settlements
Water pollution (untreated sewage into the Ganga from cities such as Varanasi)	■ Leaking sewers, landfill and industrial waste ■ In some LEDCs, agricultural pollution from fertilisers and manure is a problem	■ Control of point sources of pollution at source by regulation and fining; development of mains drainage systems and sewers ■ Removal of contaminated land sites
Water supply (overuse of groundwater led to subsidence and flooding in Bangkok)	■ Aquifer depletion, ground subsidence and low flow of rivers	■ Construction of reservoirs, pipeline construction from long-distance catchment, desalination of salt water ■ Water conservation strategies
Transport-related issues (average speed of traffic in São Paulo is 3 km/hour)	■ Rising vehicle ownership leads to congestion, noise pollution, accidents and ill-health due to release of carbon monoxide, nitrogen oxide and, indirectly, low-level ozone ■ Photochemical smog formation closely related to urban sprawl	■ Introduction of cleaner car technology (unleaded petrol catalytic converters); monitoring and guidelines for various pollution levels; movement from private car to public transport; green transport planning ■ Creation of compact and more sustainable cities

Figure 38 Environmental problems in urban areas

Figure 39 Environmental problems in an urban area – closely packed housing in Seoul, Korea and an open sewer, Castries, St Lucia

The growth of out-of-town shopping centres

Shopping in MEDCs and NICs has changed from an industry dominated by small firms to one being led by large companies. The retailing revolution has focused on superstores, hypermarkets and out-of-town shopping centres (Figure 40). These are located on 'greenfield' suburban sites with good accessibility and plenty of space for parking and future expansion. The increasing use of out-of-town shopping centres, and the trend for less frequent shopping, has led to the closure of many small shops which relied on regular sales of daily items.

Figure 40 Inside and outside the Giant hypermarket, Bandar Seri Begawan, Brunei

Advantages	Disadvantages
Plenty of free parking	They destroy large amounts of undeveloped greenfield sites
Lots of space so shops are not cramped	They destroy valuable habitats
New developments so usually quite attractive	They lead to pollution and environmental problems at the edge of town
Easily accessible by car	An increase in impermeable surfaces (shops, car parks, roads, etc.) may lead to an increase in flooding and a decrease in water quality
Being large means the shops can sell large volumes of goods and often at slightly lower prices	They only help those with cars (or those lucky enough to live on the route of a courtesy bus) – people who do not benefit might include the elderly, those without a car, those who cannot drive
Having a large shop means that individual shops can offer a greater range of goods than smaller shops	Successful out-of-town developments may take trade away from city centres and lead to a decline in sales in the CBD
Being on the edge of town means the land price is lower so the cost of development is kept down	Small businesses and family firms may not be able to compete with the vast multinational companies that dominate out-of-town developments – there may be a loss of the 'personal touch'
Developments on the edge of town reduce the environmental pressures and problems in city centres	They cause congestion in out-of-town areas
Many new jobs may be created both in the short term (construction industry) and in the long term (retail industry and linked industries such as transport, warehousing, storage, catering, etc.)	Many of the jobs created are unskilled

Figure 41 Advantages and disadvantages of out-of-town shopping centres

Figure 42 Site of an out-of-town shopping centre, Kidlington, UK

The changes in retailing have been brought about by:
- suburbanisation of more affluent households
- technological change, for example, more families own deep freezers
- economic change, with increased standards of living, especially car ownership
- traffic congestion and inflated land prices in city centres
- social changes, such as more working women.

The initial out-of-town developments came in the late 1960s and early 1970s. Now more than 20 per cent of shopping expenditure in MEDCs takes place in out-of-town stores.

Activity

Study Figure 42.

a) What are the advantages of this site for the supermarket?

b) Which population groups benefit from out-of-town developments such as the one shown in the photograph? Give reasons for your choice.

The Natural Environment

UNIT
1

Plate tectonics

Structure, landforms and landscape processes

Structure

Plate tectonics is a set of ideas that describes and explains the global distribution of earthquakes, volcanoes, fold mountains and rift valleys (Figure 1). The cause of earth movement is huge convection currents in the Earth's interior, which rise towards the Earth's surface, drag continents apart and cause them to collide. These happen because the Earth's interior consists of semi-molten layers (magma) and the Earth's surface or crust (plates) moves around on the magma. There are seven large plates (five of which carry continents) and a number of smaller plates. The main plates are the Pacific, Indo-Australian, Antarctic, North American, South American, African and Eurasian plates. Smaller ones include the Caribbean, Iranian, Arabian and Juan de Fuca plates. These move relative to one another and when they collide create tectonic activity and new landforms.

The structure of the Earth

There are four main layers within the Earth (Figure 2, page 56):

- The inner core is solid. It is five times more dense than surface rocks.
- The outer core is semi-molten.
- The mantle is semi-molten and about 2900 km thick.
- The crust is a solid and is divided into two main types: oceanic crust and continental crust. The depth of the crust varies between 10 km and 70 km. Continental crust is mostly formed of granite. It is less dense than the oceanic crust. Because it is denser the oceanic crust plunges beneath the continental one when they come together.

The **distribution** of the world's volcanoes and earthquakes is very uneven (Figure 3, page 56). They are mostly along plate boundaries which are regions of crustal instability and tectonic activity. There are over 1300 active volcanoes in the world, many of them under the ocean, and three-quarters of the world's active volcanoes are located in the

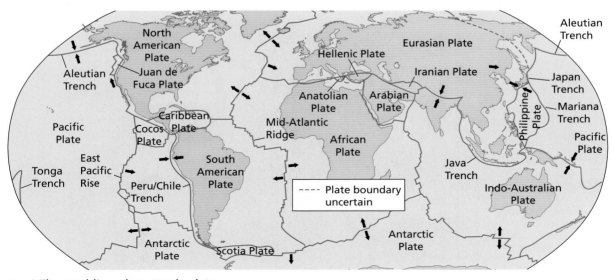

Figure 1 The world's main tectonic plates

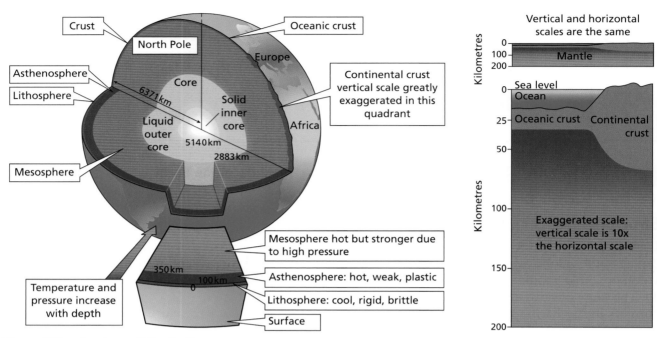

Figure 2 The structure of the Earth

Pacific Ring of Fire. Good examples include Mt Pinatubo (Philippines), Krakatoa (Indonesia) and Popcatapetl (Mexico). These volcanoes are related to plate boundaries, notably destructive plate boundaries (for example Mt St Helens in the USA and Soufrière in Montserrat in the Caribbean) and constructive ones (for example Eldfell volcano on Heimaey, Iceland). The continuing eruption of Soufrière (Montserrat) occurs at the boundary of the North American and Caribbean plates. Some volcanoes, such as Mauna Loa and Kilauea in Hawaii, and Teidi on Tenerife, are located over **hot**

spots. These are isolated plumes of rising magma that have burned through the crust to create active volcanoes.

About 500,000 earthquakes are detected each year by sensitive instruments. Most of the world's earthquakes occur in linear chains (such as along the west coast of South America) along all types of plate boundaries. Some earthquakes appear in areas away from plate boundaries such as in the mid-west of the USA. These earthquakes could still be related to plate movement as the North American plate is moving westwards. Some earthquakes are the result of

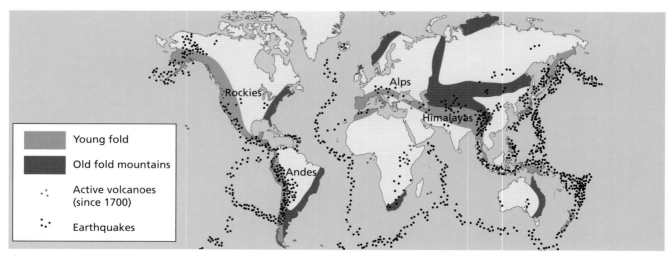

Figure 3 The distribution of fold mountains, earthquakes and volcanoes

human activity. The building of large dams and deep reservoirs increases pressure on the ground. Mining removes underground rocks and minerals and may cause collapse or subsidence of the overlying materials. Testing of nuclear weapons underground has been known to trigger earthquakes, too.

The world's main fold mountains (Rockies, Himalayas, Andes and Atlas Mountains) are related to the main collision zones – destructive plate margins and collision margins (Figure 3). In contrast, some of the older fold mountains such as the Urals in Russia, the Appalachians in North America, and the Drakensberg in South Africa, are related to folding between 250 and 450 million years ago when the location of plate boundaries was different from that of today.

Activities

1 Name the type of plate boundary located:

   a) off the west coast of central America
   b) in the south Atlantic Ocean
   c) where the Turkish plate meets the Aegean plate.

2 Wherabouts in the world is plate movement most rapid?

3 Describe the distribution of young fold mountains.

◼ Plate movement and boundaries

There are a number of different types of plate boundaries (Figure 4). These include constructive boundaries in which new oceanic crust is being created; destructive boundaries in which older crust is destroyed; collision zones where plates are folded and crumpled; and conservative plates where plates slip past each other, causing earthquakes to occur. Different plate boundaries are associated with different tectonic activities: eruptions, folding, earthquake activity, sea-floor spreading (Figure 5, page 58).

Activities

1 Study Figure 4.

   a) Describe what happens at a subduction zone.
   b) At what types of plate boundaries are volcanoes likely to occur?
   c) Which types of plate boundaries produce fold mountains?

2 Study Figure 5 (page 58) which shows a variety of tectonic landscapes.

   a) Describe the general appearance of the land in photo a). Suggest how it may have been formed.
   b) Photo b) shows a rift valley at Thingvellir in Iceland. At which type of plate boundary are rift valleys found? How might they be formed?
   c) Photo c) shows a volcanic eruption of Soufrière, Montserrat with the former capital city Plymouth in the foreground. Suggest the likely hazards of living close to a volcano.
   d) Photo d) shows tourists at the boiling mud springs at Soufrière, St Lucia. Suggest some of the advantages of living in a tectonically active region.
   e) Suggest why the volcano on Montserrat and the mud springs in St Lucia have the same name: Soufrière. What does this tell us about the processes involved in these tectonic boundaries?

Constructive margins (spreading or divergent plates)	Two plates move apart from each other causing sea-floor spreading; new oceanic crust is formed, creating mid-ocean ridges; volcanic activity is common. Mid-Atlantic Ridge (Europe is moving away from North America)
Destructive margins (subduction zones)	The oceanic crust moves towards the continental crust and sinks beneath it due to its greater density; deep sea trenches and island arcs are formed; the continental crust is folded into fold mountains; volcanic activity is common. Nazca plate sinks under the South American plate
Collision zones	Two continental crusts collide; as neither can sink they are folded up into fold mountains. The Indian plate collided with the Eurasian plate to form the Himalayas
Conservative margins (passive margins or transform plates)	Two plates slip sideways past each other but land is neither destroyed nor created. San Andreas fault in California

Figure 4 Types of plate boundaries

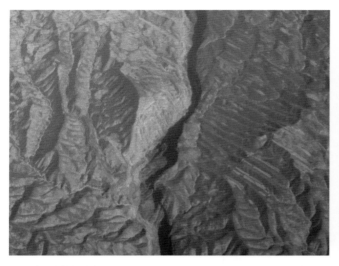

Figure 5 Tectonic activities

a Folded landscape, Himalaya foothills
b Thingvellir rift valley, Iceland
c Volcanic eruption of Soufrière, Montserrat with the former capital city Plymouth in the foreground
d Tourists standing by the boiling mud springs, Soufrière, St Lucia

Volcanoes

A volcano is an opening in the Earth's crust through which hot molten magma (lava), molten rock and ash are erupted onto the land. Most volcanoes are found at plate boundaries although there are some exceptions, such as the volcanoes of Hawaii. Some eruptions let out so much material that the world's climate is affected for a number of years. **Magma** refers to molten materials inside the Earth's interior. When the molten material is ejected at the Earth's surface through a volcano or a crack at the surface, it is called **lava**.

Key facts about volcanoes

- The greatest volcanic eruption was Tambora in Indonesia in 1815. Some 50–80 km^3 of material was blasted into the atmosphere.

- In 1883 the explosion of Krakatoa was heard from as far away as 4776 km!

- The largest active volcano is Mauna Loa in Hawaii, 120 km long and over 100 km wide.

Types of volcano

The shape of a volcano depends on the type of lava it contains. Very hot, runny lava produces gently sloping **shield volcanoes**, while thick material produces **cone-shaped volcanoes** (Figure 6). These may be the result of many volcanic eruptions over a long period of time. Part of the volcano may be blasted away during eruption. The shape of the volcano also depends on the amount of change there has been since the volcanic eruption. Cone volcanoes are associated with destructive plate boundaries, whereas shield volcanoes are characteristic of constructive boundaries and hot spots (areas of weakness within the middle of a plate).

The **chamber** refers to the reservoir of magma located deep inside the volcano. A **crater** is the depression at the top of a volcano following a volcanic eruption. It may contain a lake. A **vent** is the channel which allows magma within the volcano to reach the surface in a volcanic eruption.

Active volcanoes have erupted in recent times, such as Mt Pinatubo in 1991 and Montserrat in 1997, and could erupt again. **Dormant volcanoes** are volcanoes that have not erupted for many centuries but may erupt again, such as Mt Rainier in the USA. **Extinct volcanoes** are not expected to erupt again. Kilamanjaro in Kenya is an excellent example of an extinct volcano.

Volcanic eruptions eject many different types of materials. **Pyroclastic flows** are superhot (700 °C) flows of ash and pumice (volcanic rock) at speeds of over 500 km/h. In contrast, **ash** is very fine grained but very sharp volcanic material. **Cinders** are small rocks and coarse volcanic materials. The volume of material ejected varies considerably from volcano to volcano (Figure 7).

Eruption	Date	Volume of material ejected (km³)
Mt St Helens, USA	1980	1
Vesuvius, Italy	AD79	3
Mt Katmai, USA	1912	12
Krakatoa, Indonesia	1883	18
Tambora, Indonesia	1815	80

Figure 7 The biggest volcanic eruptions

Volcanic strength

The strength of a volcano is measured by the Volcanic Explosive Index. This is based on the amount of material ejected in the explosion, the height of the cloud it causes, and the amount of damage caused. Any explosion above level 5 is considered to be very large and violent. So far there has never been a level 8.

Predicting volcanoes

The main methods of predicting volcanoes include:
- seismometers to record swarms of tiny earthquakes that occur as the magma rises
- chemical sensors to measure increased sulphur levels
- lasers to detect the physical swelling of the volcano
- ultrasound to monitor low-frequency waves in the magma, resulting from the surge of gas and molten rock, as happened at Mt Pinatubo, El Chichon and Mt St Helens.

Living with a volcano

People often choose to live in volcanic areas because they are useful. For example:
- Some countries, such as Iceland and the Philippines, were created by volcanic activity.
- Volcanic soils are rich, deep and fertile, and allow intensive agriculture to take place.

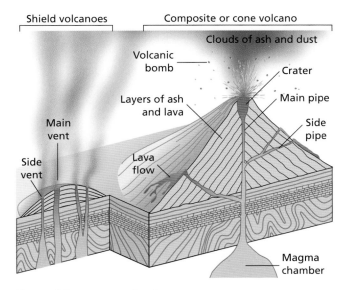

Figure 6 Two types of volcanoes

- Volcanic areas are important for tourism.
- Some volcanic areas are seen by people as being symbolic and are part of the national identity, such as Mt Fuji in Japan.

Activities

1 Describe the main characteristics of a) a shield volcano and b) a cone volcano.

2 What is the difference between an active volcano and a dormant volcano?

3 In what ways is it possible to predict volcanoes?

4 What is a pyroclastic flow? What are the dangers associated with pyroclastic flows?

Case Study

Soufrière Hills, Montserrat

Montserrat is a small island in the Caribbean, and it has been affected by a volcano since 1995. The cause of the volcano is plunging of the South American and North American plates under the Caribbean plate. Rocks at the edge of the plate melt and the rising magma forms volcanic islands.

In July 1995 the Soufrière Hills erupted after being dormant for nearly 400 years. At first the Soufrière Hills gave off clouds of ash and steam. Then in 1996 the volcano finally erupted. It caused mudflows and finally it emitted lava flows. Part of the dome collapsed, boiling rocks and ash were thrown out and a new dome was created. Ash, steam and rocks were hurled out, forcing all the inhabitants out of the south, the main agricultural part of the island (Figure 8). The largest settlement, Plymouth, with a population of just 4000, was covered in ash and abandoned (Figure 9). This has had a severe impact on Montserrat as Plymouth contained all the government offices, most of the shops and services, such as the market, post office, and cinema.

The hazard posed by the volcano was just one aspect of the risk experienced on Montserrat. For

Figure 9 Destruction of Plymouth

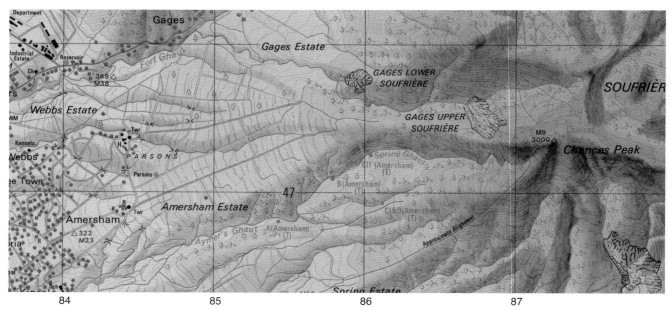

Figure 8 1:25,000 map of Soufrière and Plymouth

the refugees there were other hazards. For example, up to 50 people had to share a toilet. Sewage tanks in the temporary shelters were often not emptied for weeks on end. The risk of contamination in water and the spread of diseases, such as cholera, is greatly increased by large numbers of people living in overcrowded, unhygienic conditions.

The southern third of the island had to be evacuated. All public services (government, health, and education) had to be removed to the north of the island. Montserrat's population fell from 11,000 to 4500. Most fled to nearby Antigua. Some 'refugees' stayed on in Montserrat living in tents.

The northern part of the island has been redeveloped with new homes, hospitals, crèches, upgraded roads, ball pitch, and expansion to the island's port. The population has risen again to over 9000.

The risk of eruptions continues – scientists do not know when the current activity will cease. For now, Montserratians are learning to live with the volcano.

Activities

Study Figure 8.

1   What is the height of the highest point on the map? (The name of the highest point is Chance's Peak but this is not shown on the extract.)

2   How far is it from the Chances Peak to Sugar Bay, Plymouth (the main settlement on the map)?

3   What is the average gradient of the volcano (height/distance)? Express your answer as a 1 in x slope.

4   Describe the shape of the volcano.

5   Suggest why a settlement was built at Sugar Bay, Plymouth.

6   Outline the hazards to Plymouth as shown in Figure 9.

Case Study

Tectonic activity in Iceland

Iceland is one of the most tectonically active countries in the world. It is located at the centre of a mid-ocean ridge. This is a **constructive plate boundary**. As this new material is being forced to the Earth's surface the surrounding plates are being pushed further apart.

Iceland is a very young country and new islands such as Surtsey continue to be formed by volcanic eruptions. Over one-third of Iceland's **lava flows** are less than 10,000 years old, and the country is nowhere older than 16 million years. The north-east of Iceland is attached to the European plate and moves eastwards. In contrast, the south-west of the country is attached to the North American plate and moves westwards.

In January 1973, an eruption occurred along a 2 km long fissure. Nearly all 5300 residents of the island of Heimaey were evacuated. Strong winds blew ash and cinders from the eruption and buried homes in the main town. Massive lava flows threatened the town and the fishing port. About 70 homes and farms were buried under tephra (hardened ash) and 300 buildings were burned by fires or buried under lava flows. What makes this volcano famous is that it was the first successful attempt to manage a large eruption. For nearly six months seawater was sprayed onto the advancing lava flows in an effort to cool them down and stop them (Figure 10). In fact, the lava flows were diverted away from the town and the port and added new land (lava flows) to the island, which protected the port and made it more sheltered!

There are many advantages of tectonic activity in Iceland:

*   Natural geothermal energy – as in geysirs and hot springs. At a geyser cold water sinks into cracks and fissures and is heated under great pressure and erupts as an explosion of hot water. These are a source of geothermal energy for plants such as that at Nesjavellir (Figure 11, page 62).

Figure 10 Spraying water on the Heimaey lava flows

- It creates new land.
- It attracts tourists.

However, it does not create fertile soil in Iceland because the weather is too cold to weather the lava flows (unlike in tropical areas where new soils are soon formed).

Figure 11 Geothermal energy plant at Nesjavellir

Activities

1 Why is there tectonic activity in Iceland?

2 Suggest contrasting reasons why it was possible to manage some of the lava flows in Heimaey.

3 In what ways did the eruption of 1973 benefit the islanders of Heimaey?

4 Outline some of the advantages of volcanic activity to Iceland.

Earthquakes

Earthquakes are sudden, violent shaking of the Earth's surface. They occur after a build-up of pressure causes rocks and other materials to give way. Most of this pressure occurs at plate boundaries when one plate is moving against another. Earthquakes are associated with all types of plate boundaries. The **focus** refers to the place beneath the ground where the earthquake takes place. **Deep-focus earthquakes** are associated with subduction zones. **Shallow-focus earthquakes** are generally located along constructive boundaries and along conservative boundaries. The **epicentre** is the point on the ground surface immediately above the focus.

Some earthquakes are caused by human activity such as:

- nuclear testing
- building large dams
- drilling for oil
- coal mining.

Earthquake damage

The extent of earthquake damage is influenced by a number of factors:

- **Strength of earthquake** and **number of aftershocks** – the stronger the earthquake the more damage it can do. For example, an earthquake of 6.0 on the Richter scale is 100 times more powerful than one of 4.0; the more aftershocks there are the greater the damage that is done.
- **Population density** – an earthquake that hits an area of high population density, such as in the Tokyo region of Japan, could inflict far more damage than one that hits an area of low population and low building density.
- The **type of buildings** – MEDCs generally have better-quality buildings, more emergency services and the funds to cope with disasters. People in MEDCs are more likely to have insurance cover than those in LEDCs.
- The **time of day** – an earthquake during a busy time, such as rush hour, may cause more deaths than one at a quiet time. Industrial and commercial areas have fewer people in them on Sundays, and homes have more people in them at night.
- The **distance from the centre (epicentre)** of the earthquake – the closer a place is to the centre of the earthquake, the greater the damage that is done.
- The **type of rocks and sediments** – loose materials may act like liquid when shaken; solid rock is much safer and buildings should be built on level areas formed of solid rock.
- **Secondary hazards** – such as mudslides and tsunamis (large sea waves), fires, contaminated water, disease, hunger and hypothermia.

Dealing with earthquakes

People cope with earthquakes in a number of ways. The three basic options from which they can choose are:

- do nothing and accept the hazard
- adjust to living in a hazardous environment – strengthen your home
- leave the area.

The main ways of dealing with earthquakes include:

- better forecasting and warning
- building design, building location and emergency procedures.

There are a number of ways of predicting and monitoring earthquakes. These include:

- measuring crustal movement – small-scale movement of plates
- recording changes in electrical conductivity
- noting strange and unusual animal behaviour, for example among carp fish
- checking historic evidence – there are possibly trends in the timing of earthquakes in some regions.

Building design

A single-storey building responds quickly to earthquake forces (Figure 12). A high-rise building responds slowly, and shock waves are increased as they move up the building. If the buildings are too close together, vibrations may be amplified between buildings and increase damage. The weakest part of a building is where different elements meet. Elevated motorways are therefore vulnerable in earthquakes because they have many connecting parts.

Certain areas are very much at risk from earthquake damage – notably areas with weak rocks, faulted (broken) rocks, and soft soils. Many oil and water pipelines in tectonically active areas are built on rollers so that they can move with an earthquake rather than fracture (Figure 13).

Figure 13 Pipeline on rollers

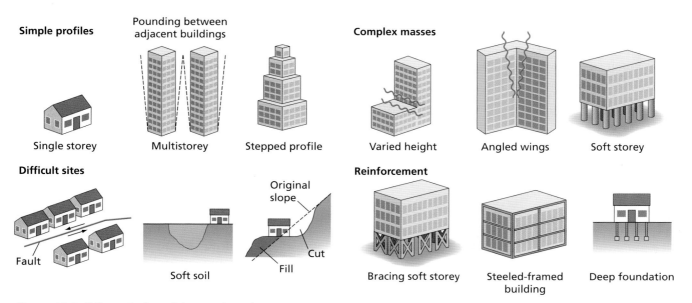

Simple profiles

Single storey

Pounding between adjacent buildings

Multistorey

Stepped profile

Complex masses

Varied height

Angled wings

Soft storey

Difficult sites

Fault

Soft soil

Original slope

Cut

Fill

Reinforcement

Bracing soft storey

Steeled-framed building

Deep foundation

Figure 12 Buildings designed for earthquakes

Activities

1 Describe the global distribution of earthquakes as shown on Figure 1 on page 55.

2 What is the difference between shallow-focus and deep-focus earthquakes?

3 Study Figure 12 (page 63). In what ways can building design reduce the impact of earthquakes?

4 Design a poster outlining the factors that influence the impact of an earthquake.

Case Study

Three earthquakes

Kobe, 1995

The Kobe earthquake of January 1995 was responsible for over 5000 deaths, 30,000 injuries and for making over 300,000 people homeless. It was caused by the oceanic Pacific plate plunging (subducting) under the continental Eurasian plate (Figure 14).

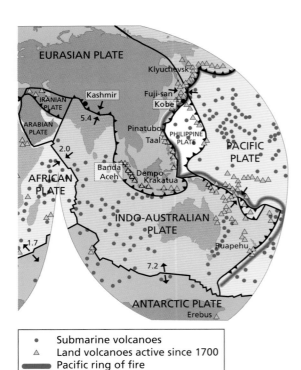

Figure 14 Plates in south and east Asia

The earthquake, which registered 7.2 on the Richter scale, struck at 5.46 in the morning. Many people were crushed in their beds although the number of people killed by collapsing motorways was relatively low. Many important buildings, such as the City Hall and public hospitals, were destroyed. Up to 80 per cent of the schools, museums and sports facilities were also destroyed.

The earthquake came as a surprise to Japanese scientists since the area was considered to be one of the safest for earthquake activity. Conditions for the survivors worsened as rain, strong winds and lightning increased the risk of landslides. Doctors were faced with outbreaks of disease due to the damp, unhygienic conditions. Over 1300 aftershocks were recorded, and these toppled many buildings. Gas and water pipes were broken and there were 175 separate fires in Kobe. Water supplies to deal with the fire were badly disrupted.

Transport and communications were badly affected. A 1 km stretch of elevated highway collapsed and Japan's bullet train was closed. The damage was estimated at $160 billion.

The south Asian tsunami, 2004

The term 'tsunami' is the Japanese for 'harbour wave'. About 90 per cent of these events occur in the Pacific Basin. Tsunami are generally caused by earthquakes (usually in subduction zones) but can also be caused by volcanoes, for example Krakatoa (1883), and landslides, for example Alaska (1964). Tsunamis have the potential to cause widespread disaster as in the case of the Boxing Day tsunami 2004. It became a global disaster, killing people from nearly 30 countries, many of them foreign tourists (Figure 15, page 65). Between 180,000 and 280,000 people were killed in the 2004 tsunami.

The cause of the tsunami was a giant earthquake and landslide caused by the sinking of the Indian plate under the Eurasian plate. Pressure had built up over many years and was released in the earthquake which reached 9.0 on the Richter scale.

The main impact of the Boxing Day tsunami was on the Indonesian island of Sumatra, the closest inhabited area to the epicentre of the earthquake (Figure 15). More than 70 per cent of the inhabitants of some coastal villages died. Apart from Indonesia, Sri Lanka suffered more from the tsunami than anywhere else – at least 31,000 people are known to have died there, mostly along the southern and eastern coastlines.

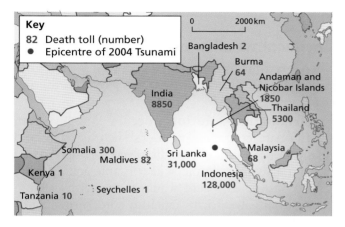

Figure 15 Death toll in the south Asian tsunami

Kashmir 2005

In autumn 2005 Kashmir was struck by an earthquake which recorded 7.7 on the Richter scale. There were over 22 aftershocks in the 24 hours after the main earthquake, some of which measured over 6 on the Richter scale. The quake and its aftershocks were felt from central Afghanistan to western Bangladesh. The cause of the earthquake was the Indian plate (Indo-Australian plate) moving against the Eurasian plate and the Iranian plate.

Buildings were wrecked in an area spanning at least 400 km from Jalalabad in Afghanistan to Srinagar in Indian Kashmir. The death toll was over 73,000. More than 3 million people were left homeless and 69,000 seriously injured. Of the homeless, nearly 1 million had to sleep in the open. In built-up areas, water and sanitation systems were broken, giving a high risk of an outbreak of disease. There was a desperate shortage of tents capable of withstanding the Kashmiri winter, of blankets, sleeping-bags and warm clothes, as well as of medicine and food.

Some charities expressed concern that the public may be suffering **compassion fatigue**. The UN had received just 12 per cent of the $312 million pledged to its emergency appeal, in contrast with 80 per cent of pledges at the same stage after the south Asian tsunami. Whatever the reason, the earthquake did not provoke the response from the rest of the world that it needed. Perhaps the tsunami, killing hundreds of thousands, deadened compassion.

Activities

1 Comment on the impacts of the 1995 earthquake in Kobe, Japan.

2 Suggest reasons why each of the three earthquakes occurred in south and east Asia.

3 Why was the 2004 tsunami considered to be a 'global disaster'?

4 How and why were the impacts of the Kashmir earthquake different from those of the Kobe earthquake?

Landforms and landscape processes

Weathering

Weathering is the **decomposition** (chemical breakdown) and **disintegration** (physical breakdown) of rocks *in situ*. Decomposition refers to the chemical process which creates altered rock substances, whereas disintegration or physical weathering produces smaller, angular fragments of the same rock. Weathering is important for landscape evolution as it breaks down rock and facilitates erosion and transport. It also helps form soil which is essential for the growth of crops.

Physical (mechanical) weathering

There are four main types of physical weathering: freeze–thaw (ice crystal growth), salt crystal growth, disintegration and pressure release.

- **Freeze–thaw** occurs when water in joints and cracks freezes at $0\,°C$ and expands by 10 per cent and exerts pressure up to $2100\,kg/cm^2$. Rocks can only withstand a maximum pressure of about $500\,kg/cm^2$. It is most effective in damp environments where moisture is plentiful and there are frequent fluctuations above and below freezing point, for example in periglacial and alpine regions.
- **Salt crystal growth** occurs in two main ways: first in areas where temperatures fluctuate around $26–28\,°C$, and where water evaporates leaving salt crystals behind, as in hot deserts, to attack the structure. Sodium sulphate (Na_2SO_4) and sodium carbonate (Na_2CO_3) expand by 300 per cent. Both mechanisms are frequent in hot desert regions.
- **Disintegration** is also found in hot desert areas where there is a large diurnal temperature range. Rocks heat up by day and contract by night. As rock is a poor conductor of heat, expansion and contraction in the outer layers causes peeling or **exfoliation** to occur (Figure 1). Griggs (1936) showed that moisture is essential for this to happen.
- **Pressure release** is the process in which overlying rocks are removed by erosion thereby allowing underlying ones to expand and fracture parallel to the surface. The removal of a great weight, such as a glacier, has the same effect.

Chemical weathering

There are four main types of chemical weathering: carbonation-solution, hydrolysis, hydration and oxidation.

- **Carbonation-solution** occurs on calcium carbonate contained in rocks such as chalk and limestone. Rainfall and dissolved carbon dioxide form a weak carbonic acid (organic acids acidify water too). The acid water reacts with calcium carbonate to form calcium bicarbonate, or calcium hydrogen carbonate, which is soluble and is removed by percolating water.
- **Hydrolysis** occurs on rocks that include orthoclase feldspar, for example granite (Figure 2). Orthoclase reacts with acid water and forms kaolinite (or kaolin or china clay), silicic acid and potassium hydroxyl. The acid and hydroxyl are removed in the solution leaving china clay behind as the end product. Other minerals in the granite, such as quartz and mica, remain in the kaolin.
- **Hydration** is the process whereby certain minerals absorb water, expand and change, for example gypsum becomes anhydrate.
- **Oxidation** occurs when iron compounds react with oxygen to produce a reddish-brown coating (rust).

In addition to mechanical and chemical weathering, a third type is sometimes included, **biological weathering**. This is the action of plants on rocks.

Figure 1 Rock disintegration – chemically weathered granite

Figure 2 Weathering of granite near Bulawayo, Zimbabwe

Figure 3 Biological weathering on a wall

There are two main impacts:

- mechanical – the prising of rocks apart by the growth of roots
- chemical – the breakdown of rocks by the release of organic acids from the plants onto the rock surface (Figure 3).

Activities

1 Why is weathering important?

2 In what ways does chemical weathering differ from mechanical weathering?

3 Describe one type of physical weathering and its results.

4 Describe one type of chemical weathering and its results.

5 Suggest how the plants in Figure 3 are weathering the surface they are on.

6 Explain how the rock is being weathered in Figure 1.

Controls of weathering

The type and rate of weathering varies with climate (Figure 4, page 68). Peltier's diagrams (1950) show how weathering is related to moisture availability and average annual temperature. In general, frost shattering increases as the number of freeze–thaw cycles increases. By contrast, chemical weathering increases with moisture and heat. According to **Van't Hoff's Law**, the rate of chemical weathering increases 2–3 times for every increase of temperature of 10 °C (up to a maximum temperature of 60 °C). The efficiency of freeze–thaw, salt crystal growth and insolation weathering is influenced by:

- critical temperature changes
- frequency of cycles
- diurnal and seasonal variations in temperature.

Geology

Rock type and rock structure influence the rate and type of weathering in many ways, due to:

- chemical composition
- the nature of cements in sedimentary rock
- joints and bedding planes.

For example, limestone consists of calcium carbonate and is therefore susceptible to carbonation-solution. By contrast granite is prone to hydrolysis because of

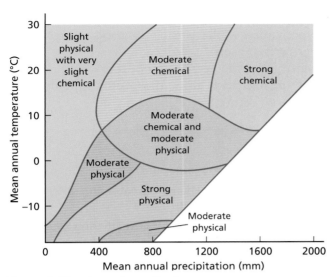

Figure 4 Climate and weathering

the presence of feldspar. In sedimentary rocks, the nature of the cement is crucial. Iron-oxide based cements are prone to oxidation.

Joint patterns exert a strong control on water movement. Where there is a greater density of joints, rocks are likely to be weathered more (Figure 5). Where there are fewer joints there is less opportunity

to weather the rock. Similarly, grain size influences the speed at which rocks weather. Coarse-grained rocks weather quickly owing to high permeability. On the other hand, fine-grained rocks offer a greater surface area for weathering and may be highly susceptible to weathering. The type of mineral that forms a rock is important too. Rocks formed of resistant minerals, such as quartz, muscovite and feldspar in granite, will resist weathering. By contrast, rocks formed of weaker minerals will weather rapidly. The interrelationship of geology and climate on the development of landforms is well illustrated by limestone and granite.

Activities

1 Study Figure 4 which shows the relationship of climate and type and intensity of weathering.

   a) Under what conditions does strong chemical weathering occur?
   b) What type and intensity of weathering occurs when the mean annual temperature is 5 °C and the mean annual precipitation is 1000 mm?
   c) Under what conditions does strong physical weathering take place?
   d) Briefly explain how freeze–thaw weathering takes place.

2 Figure 5 shows how rock characteristics may influence weathering.

   a) Briefly explain two ways in which rock type can influence the type and intensity of weathering.
   b) Apart from climate and rock type, suggest one factor that affects the type and rate of weathering in an area. Explain how it does so.

Surface

Original surface
Breakdown of rock along joints and bedding planes

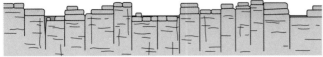

Bedding planes

Original surface
Removal of weathered material to expose tors

Figure 5 Joint pattern and weathering: formation of tors

Landforms associated with limestone

Limestone is a hard rock and a permeable one. Being permeable it allows water to percolate (seep) into it, along lines of weaknesses known as joints and bedding planes. This helps create many distinctive landforms. The main form of weathering to affect limestone is carbonation-solution. The effectiveness of solution is related to the pH of the water. The more acidic the water the greater the rate of carbonation-solution.

There are important differences between limestone in tropical and subtropical areas, and limestone in cooler, temperate areas. Limestones in tropical and subtropical areas often show rounded landforms, known as cockpit karst (Figure 6) and tower karst (Figure 7) while those in temperate areas may be more angular and rugged. Rates of weathering are often greater in tropical areas because:

- higher temperatures speed up chemical reactions
- there is a plentiful supply of moisture in many tropical areas, especially rainforests
- vegetation releases organic acids which makes water in the soil more acidic.

Limestone has a distinctive bedding plane and joint pattern, which is described as being **massively jointed**. This means that the joints are often spaced quite far apart, often at regular intervals. The joints are weaknesses that allow water to percolate into the rock and dissolve it.

Surface features

The term **karst** refers to well-developed features on dry limestone – that is, without surface drainage. The joints and cracks are attacked and enlarged over thousands of years. **Clints** (outcrops of bare rock) and **grikes** (the gaps between the rocks where the joints are located) develop on the surface of the exposed limestone. Large areas of bare exposed limestone are known as **limestone pavements**.

Dolines are large depressions formed by the solution or collapse of limestone. Depressions can range from small-scale sinks to large **uvaalas** up to 30 m in diameter. **Swallow holes** (or **sinks**) are smaller depressions in the landscape, also caused by the solution of limestone. Often a river disappears down the hole, hence the term 'sink'.

Underground features

Underground features include caves and tunnels formed by carbonation-solution and erosion by rivers. Carbonation is a reversible process. When calcium-rich water drips from the ceiling it leaves behind calcium in the form of stalactites and stalagmites. **Stalactites** develop from the top of the cave whereas **stalagmites** are formed on the base of the cave. Rates of deposition are slow, about 1 mm in 100 years.

Features of tropical karst

There are two major landform features associated with tropical karst. Polygonal or cockpit karst (Figure 6) is a landscape pitted with smooth-sided, soil-covered depressions and cone-like hills. Tower karst (Figure 7) is a landscape characterised by upstanding rounded blocks set in a region of low relief.

Polygonal or **cockpit karst** is characterised by groups of hills, fairly uniform in height. These can be up to 160 m high in Jamaica, with a base of up to 300 m. They develop mainly as a result of solution. By contrast, **tower karst** is much more variable in size than the conical hills of cockpit karst, and ranges from just a few metres to over 150 m in height in Sarawak. Other areas of tower karst include southern China, Malaysia and Indonesia. They are characterised by steep sides, with cliffs and overhangs, and with caves and solution notches at their base.

Figure 6 Cockpit karst, South Africa

Figure 7 Tower karst, China

Granite

Granite is an igneous, crystalline rock. It has great physical strength and is very resistant to erosion. There are many types of granite but all share certain characteristics. They contain quartz, mica and feldspar. These are resistant minerals. The main processes of weathering that occur on granite are freeze–thaw and hydrolysis.

Tors are isolated masses of bare rock (Figure 8). Some of the boulders of the mass are attached to part of the bedrock. Others merely rest on the top. **Bornhardts** are large, isolated, dome-like hills standing above an extensive plain. They are often made of granite and formed by exfoliation weathering. They are common in savanna areas.

Granite's resistance to weathering results in only a thin, gritty soil cover. Such soils are generally infertile, so rough grazing is the dominant land use in granite areas. It is an impermeable rock and many marshy hollows at the heads of the valleys indicate the limited downward movement of water.

Figure 8 A distinctive tor: The Holy Family, Rhodes Memorial, Matopas, Zimbabwe

Activities

1 What is the difference between a *clint* and a *grike*?

2 What is another name for a *sink*?

3 What is the difference between *cockpit karst* and *tower karst*?

4 Why is there so much carbonation in tropical areas?

River processes

Erosion

The main types of erosion include:

- **abrasion** (or **corrasion**) – the wearing away of the bed and bank by the load carried by a river
- **attrition** – the wearing away of the load carried by a river, which creates smaller, rounder particles
- **hydraulic action** – the force of air and water on the sides of rivers and in cracks
- **solution** (or **corrosion**) – the removal of chemical ions, especially calcium, which causes rocks to dissolve.

There are many factors affecting erosion. These include:

- **load** – the heavier and sharper the load the greater the potential for erosion
- **velocity and discharge** – the greater the velocity and discharge the greater the potential for erosion
- **gradient** – increased gradient increases the rate of erosion
- **geology** – soft, unconsolidated rocks, such as sand and gravel, are easily eroded
- **pH** – rates of solution are increased when the water is more acidic
- **human impact** – deforestation, dams, and bridges interfere with the natural flow of a river and frequently end up increasing the rate of erosion.

Transportation

The main types of transportation in a river (Figure 9) include:

- **suspension** – small particles are held up by turbulent flow in the river
- **saltation** – heavier particles are bounced or bumped along the bed of the river
- **solution** – the chemical load is carried dissolved in the water

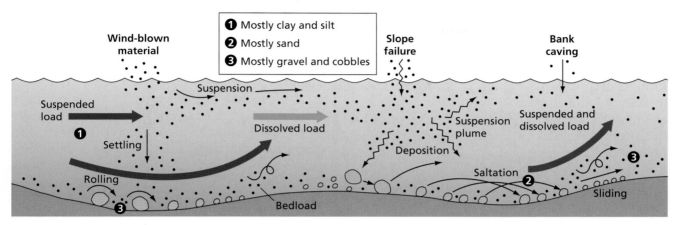

Figure 9 Types of transport in a river

- **traction** – the heaviest material is dragged or rolled along the bed of the river
- **flotation** – leaves and twigs are carried on the surface of the river.

Deposition

Deposition occurs as a river slows down and it loses its energy. Typically, this occurs as a river floods across a floodplain, or enters the sea, or behind a dam. It is also more likely during low flow conditions (such as in a drought) than during high flow (flood) conditions – as long as the river is carrying sediment. The larger, heavier particles are deposited first, the smaller, lighter ones later. Features of deposition include deltas, levées, slip-off slopes (point bars), ox-bow lakes, braided channels, and floodplains.

Downstream changes

Rivers have three main roles: to erode the river channel, to transport materials, and to create new erosional and depositional landforms. Most rivers have three main zones – a zone of erosion, a zone of transport, and a zone of deposition. Erosion, transport and deposition are found in all parts of a river, though one process tends to be dominant. For example, there is more erosion in the upper part, while there is more deposition in the lower course. This is related to the changes in a river downstream. Figure 10 shows that velocity, discharge and load increase downstream whereas gradient and the size of load decrease downstream.

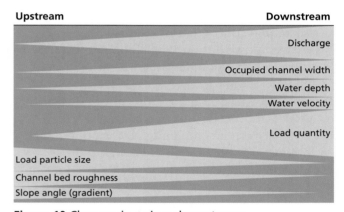

Figure 10 Changes in a river downstream

Activities

1 a) Briefly describe the four main ways in which rivers erode.
  b) Suggest how they will vary with **i)** velocity of water **ii)** rock type and **iii)** pH of water.

2 a) What are the main types of transport?
  b) How might the type and quantity of the river's load vary between flood conditions and low flow conditions?

3 Study Figure 10.

  a) Describe how the amount and size of load varies downstream.
  b) Suggest reasons for the changes you have identified.
  c) State a reason why the channel bed roughness decreases downstream.
  d) How might the nature of the load affect the type and amount of erosion carried out by the river? Give reasons for your answer.

■ Landscapes of river erosion

The long profile

The longitudinal section through a river's course, from its source to its mouth, is known as a **long profile**. A number of processes, such as weathering and mass movement, interact to create variations in the profile (Figure 11). Irregularities, or **knick points**, may be due to:

- geological structure, for example hard rocks erode slowly, which can result in the formation of waterfalls and rapids
- variations in the load, for example when a tributary with a coarse load may lead to a steepening of the gradient of the main valley
- sea-level changes – relative fall in sea level (isostatic recovery, eustatic fall, etc.) will lead to renewed downcutting which enables the river to erode former floodplains and form new terraces and knick points.

Rivers tend to achieve a condition of equilibrium, or grade, and erode the irregularities. There is a balance between erosion and deposition in which a river adjusts to its capacity and the amount of work being done. The main adjustments are in channel gradient leading to a smooth concave profile.

Cross profiles

The **cross profile**, or cross-section, of the upper part of a river is often described as V-shaped (Figure 11). Rivers in their upper course typically have a steep gradient and a narrow valley. The rivers are shallow and fast-flowing. There is normally much friction with large boulders and much energy is used to overcome friction. The processes likely to occur are vertical erosion, weathering on the slopes, mass movements and transportation. Features likely to be found include waterfalls, rapids, potholes, gorges and interlocking spurs.

In the middle course of the river, the valley is still V-shaped but is less steep. Slopes are gentler. A floodplain is beginning to form and meanders are clear. Processes in the middle course include erosion (both vertical and lateral), meandering, transport, and some deposition on the inner bends of the meanders.

In contrast, in the lower course the cross profile is much flatter. Processes include erosion (on the outer banks), transport, and deposition (especially on the inner bends and on the floodplain). Characteristic features include levées, ox-bow lakes, floodplains, deltas and terraces.

Features of river erosion

Localised erosion by hydraulic action and abrasion, especially by large pieces of debris, may lead to the formation of **potholes**. These are typically seen in the upper course of a river when the load is larger and more rugged. **Waterfalls** frequently occur on

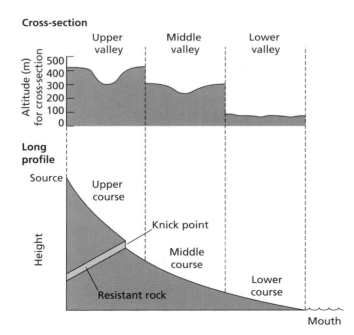

Figure 11 Long and cross profiles

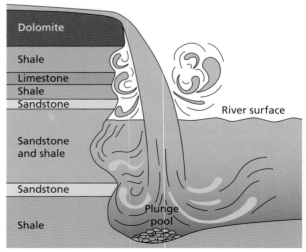

Figure 12 Erosion by a waterfall

horizontally bedded rocks. The soft rock is undercut by hydraulic action and abrasion. The weight of the water and the lack of support cause the waterfall to collapse and retreat (Figure 12). Over thousands of years the waterfall may retreat enough to form a **gorge of recession** (Figure 13). Where there are small outcrops of hard and soft rock, **rapids** may develop rather than a waterfall.

Ox-bow lakes are the result of erosion and deposition. Lateral erosion, caused by fast flow in the **meanders**, is concentrated on the outer, deeper bank of a meander. During times of flooding, erosion increases. The river breaks through and creates a new, steeper channel. In time, the old meander is closed off by deposition to form an ox-bow lake.

Rivers typically meander. This means that the water does not follow a straight line but takes a curving route. As a result there are variations in the water's speed across a river. Velocity is fastest on the outside bank and slowest on the inside bank. There is therefore erosion on the outer bank and deposition on the inner bank. This produces steep river cliffs on the outer bank of a meander and gentle slip-off slope on the inner bank.

Figure 13 1:25,000 map of the Niagara Falls area

Activity
Refer to Figure 13 (page 73) for this activity.

1 Which two countries share a border at Niagara Falls?

2 In which direction is the river flowing?

3 What is the map evidence that there is a gorge below Niagara Falls?

4 Using map evidence, suggest how the Niagara River has been used for human activities.

5 Approximately how wide is Niagara Falls in grid square 5771?

■ Landscapes of river deposition

There are many features associated with deposition by rivers. These include floodplains, levées, braided channels and deltas (Figure 14).

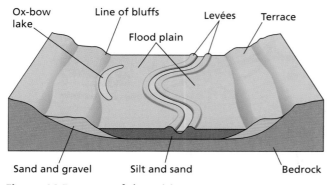

Figure 14 Features of deposition

Floodplains

The area covered by water when a river floods is known as its **floodplain**. When a river's discharge exceeds the capacity of the channel, water rises over the river banks and floods the surrounding low-lying area. Sometimes a floodplain will itself be eroded following a fall in sea level. When this happens, the remnants of the old floodplain are left behind as river terraces. These are useful for settlement as they are above the new level of the floodplain and are generally free from flooding.

Levées

When a river floods its speed is reduced, slowed down by friction caused by contact with the floodplain. As its velocity is reduced the river has to deposit some of its load. It drops the coarser, heavier material first to form raised banks, or **levées**, at the edge of the river. This means that over centuries the levées are built up of coarse material, such as sand and gravel, while the floodplain consists of fine silt and clay.

Braided channels

Braiding occurs when a river transports a very heavy load in relation to its velocity. If a river's discharge falls its competence and capacity are reduced. This forces the river to deposit large amounts of its load and multi-channels, or **braided channels**, are formed. It is common in rivers that experience seasonal variations in discharge. For example, in proglacial and periglacial areas such as southern Iceland most of the discharge occurs in late spring and early summer, as snow and ice melt. This enables river to carry very large loads which are quickly deposited as discharge decreases.

Deltas

For deltas to be formed a river needs to:

- carry a large volume of sediment, for example rivers in semi-arid regions and in areas of intense human activity
- enter a still body of water which causes velocity to fall, the water loses its capacity and competence and deposition occurs, with the heaviest particles being deposited first and the lightest last.

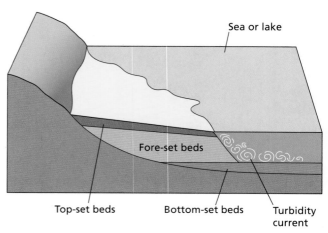

Figure 15 Formation of a delta

Deposition is increased if the water is salty, as this causes salt particles to group together, become heavier, and be deposited. Vegetation also increases the rate of deposition by slowing down the water.

Activities

1 In your own words describe the following:

   a) a floodplain
   b) an ox-bow lake
   c) a levée.

2 Briefly explain how levées are formed.

3 How are floodplains useful for human activity?

■ Marine processes

The factors that affect coastal processes and coastal landforms include:

- waves and currents, including longshore drift
- local geology – that is, rock type, structure and strength
- changes in sea level
- human activity and the increased use of coastal engineering.

All of these factors interact, and produce a unique set of processes that occur at the coast. These processes go on to produce different types of landform for every coastal area.

Wave refraction and longshore drift

Waves result from friction between wind and the sea surface. Waves in the open, deep sea are different from those breaking on shore. Sea waves are forward-moving surges of energy. Although the shape of the surface wave appears to move, the water particles follow a roughly circular path within the wave. As waves approach the shore, their speed is reduced as they touch the sea floor (Figure 16a). Wave **refraction** causes two main changes: the speed of the wave is reduced, and the shape of the wave front is altered. If refraction is completed, the wave fronts will break parallel to the shore (Figure 16b).

Wave refraction also distributes wave energy along a stretch of coast. On a coastline with alternating headlands and bays, wave refraction concentrates destructive/erosive activity on the headlands, while deposition tends to occur in the bays. Irregularities in the shape of the coastline mean that refraction is not always totally achieved. This causes **longshore drift**, which is a major force in the transport of material along the coast (Figure 16c). It occurs when waves move up to the beach (this is known as **swash**) in one direction, but the waves draining back down the beach (known as **backwash**) take a different route (under the effect of gravity). The net movement is *along* the shore – hence the term longshore drift. A wooden or concrete wall (groyne) may be built to prevent longshore drift from removing sand or shingle from the beach.

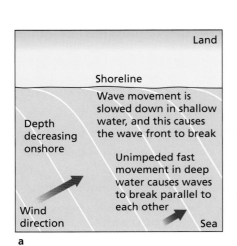

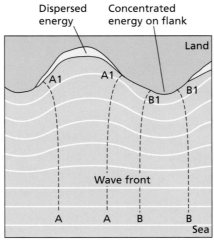

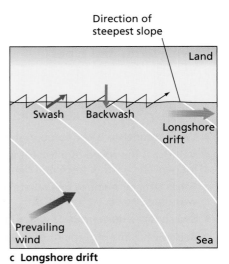

Figure 16 Wave refraction and longshore drift

Case Study

Human activity and longshore drift in West Africa

In Ghana the increase in coastal retreat has been blamed on the construction of the Akosombo Dam on the Volta River. It is just 110 km from the coast and disrupts the flow of sediment from the Volta River and stops it from reaching the shore. Thus there is less sand to replace that which has already been washed away by longshore drift, and so the coastline retreats due to erosion by the Guinea Current. Towns such as Keta, 30 km east of the Volta estuary, have been destroyed as their protective beach has been removed.

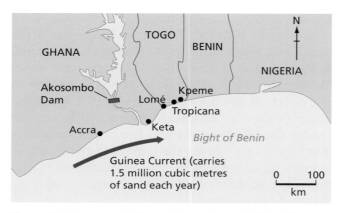

Figure 17 Human activity and longshore drift in West Africa

Waves: terms and types

- **Wavelength** is the distance between two successive crests or troughs.
- **Wave height** is the distance between the trough and the crest.
- **Wave frequency** is the number of waves per minute.
- **Velocity** is the speed of a travelling wave, and is influenced by wind, fetch and depth of water.
- **Fetch** is the amount of open water over which a wave has passed.
- **Swash** is the movement of water up the beach.
- **Backwash** is the movement of water down the beach.
- Waves are sometimes divided into **constructive waves** and **destructive waves**.

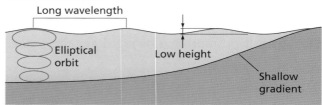

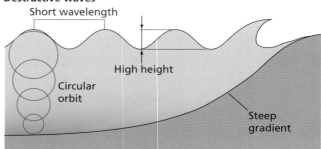

Destructive waves (erosional waves)	Constructive waves (depositional waves)
■ Short wavelength (< 20 m)	■ Long wavelength (up to 100 m)
■ High height (> 1 m)	■ Low height (< 1 m)
■ High frequency (10–12/minute)	■ Low frequency (6–8/minute)
■ Low period (one every 5–6 seconds)	■ High period (one every 8–10 seconds)
■ Backwash > swash	■ Swash > backwash
■ Steep gradient	■ Low gradient
■ Caused by local winds and storms	■ Caused by swell from distant storms
■ High-energy waves	■ Low-energy waves

Figure 18 Destructive and constructive waves

Activities

1 Define the following terms: *swash, fetch, wave refraction, longshore drift, backwash.*

2 Describe the main differences between a destructive wave and a constructive wave.

3 Describe and explain the process of longshore drift.

4 Briefly describe how human activity has affected the impact of longshore drift in West Africa.

Landscapes of marine erosion

Many types of erosion are carried out by waves:

- **Hydraulic action** occurs as waves hit or break against a cliff face. Any air trapped in cracks is put under great pressure. As the wave retreats, this build-up of pressure is released with explosive force. This is especially important in well-jointed rocks such as limestone, sandstone and granite, and in weak rocks such as clays and glacial deposits. Hydraulic action makes the most impact during storms.
- **Abrasion** is the process of a breaking wave hurling materials, such as pebbles or shingle, against a cliff face. It is similar to abrasion in a river.
- **Attrition** is the process in which eroded material, such as broken rock, is worn down to form smaller, rounder beach material.
- **Solution** occurs on limestone and chalk. Calcium carbonate, a salt found in these rocks, dissolves slowly in acidic water.

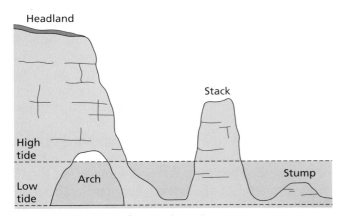

Figure 19 Features of coastal erosion

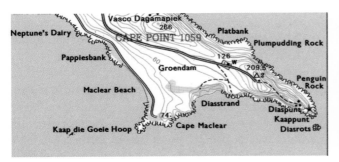

Figure 20 The Cape Peninsula, South Africa – photo shows Cape Maclear and Cape of Good Hope in the background

Features of marine erosion

On a headland, erosion will exploit any weakness, creating, at first, a **cave**. Once the cave reaches both sides of the headland, an **arch** is formed. A collapse of the top of the arch forms a **stack**, and when the stack is eroded a **stump** is created (Figure 19). Where erosion opens up a vertical crack, allowing sea water to spout up at the surface, a **blowhole** is formed. The sandstone of the Cape Peninsula in South Africa has been attacked by the sea forming steep vertical cliffs, and small-scale features such as arches and stacks (Figure 20).

Wave action is concentrated between the high water mark (HWM) and the low water mark (LWM). HWM is the highest level reached by the sea at high tide; LWM is the lowest level reached by the sea at low tide. It may undercut a cliff face, creating a notch and overhang (Figure 21, page 78). As erosion continues, the notch becomes deeper and eventually the overhang collapses, causing the cliff line to retreat. The base of the cliff is left behind as an increasingly longer platform. This is sometimes called a **wave-cut platform**, because it has been cut or eroded by wave action.

Cliff profiles vary greatly, depending on the:

- rate of coastal erosion (cliff retreat)
- strength of the rock
- presence of joints and bedding planes.

In addition, cliffs change over time, from ones dominated by marine processes (like wave erosion) to ones protected from marine processes but affected by land-based processes.

On a larger scale, bays may be eroded in beds of weaker rock (Figure 22). The harder rocks form headlands that protrude whereas the weaker rocks are eroded to form bays. Wave refraction in the bay spreads wave energy around the bay and focuses wave energy on the flanks of the headlands. Bayhead beaches are formed when constructive waves deposit sand between two headlands, such as at Maracas Bay and Tyrico Bay in northern Trinidad.

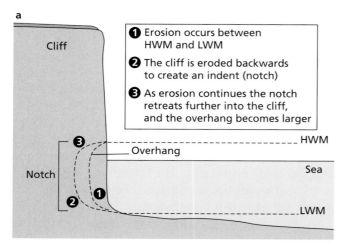

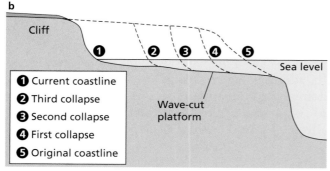

Figure 21 Formation of a wave-cut platform

Figure 22 Headlands and a bay: Praia de Rocha, Portugal

Activities

1 What is the difference between attrition and erosion?

2 Why does hydraulic action occur on jointed rocks?

3 What type of rocks are affected by solution?

4 What types of erosion are most likely to take place:

   a) during a storm
   b) on beaches
   c) on the face of a cliff?

5 In your own words, describe how a wavecut platform may be formed.

6 Make a sketch of Figure 22 and label the following features: headland, bay, stack, beach, cave.

■ Landscapes of marine deposition

Beaches

Excellent beach development occurs on a lowland coast (constructive waves), with a sheltered aspect/trend, composed of 'soft' rocks (which provides a good supply of material), or where longshore drift supplies abundant material.

The term **beach** refers to the accumulation of material deposited between low spring tides and the highest point reached by storm waves at high spring tides. A typical beach will have three zones: backshore, foreshore and offshore. The backshore is marked by a line of dunes or a cliff. Above the high water mark there may be a **berm** or **shingle ridge**. This is coarse material pushed up the beach by spring tides and aided by storm waves flinging material well above the level of the waves themselves. These are often referred to as **storm beaches**. The seaward edge of the berm is often scalloped and irregular due to the creation of beach **cusps**.

The foreshore is exposed at low tide. Offshore, the first (heaviest) material is deposited. In this zone, the waves touch the sea bed and so the material is usually disturbed, sometimes being pushed up as **offshore bars**, when the offshore gradient is very shallow. Offshore bars are usually composed of coarse sand or shingle. Between the bar and shore, lagoons (often called **sounds**) develop. If the lagoonal water is calm and fed by rivers, marshes and mudflats can be found. Bars can be driven onshore by storm winds and waves. A classic area is off the coast of the Carolinas in the south-east of the USA.

Bars and spits

These more localised features will develop where:

- abundant material is available, particularly shingle and sand
- the coastline is irregular due, for example, to local geological variety
- where there are estuaries and major rivers.

A **spit** is a beach of sand or shingle linked at one end to land. It is found on indented coastlines or at river mouths. For example, wave energy is reduced along a coast where headlands and bays are common and near river mouths (estuaries and rias).

Spits often become curved as waves undergo refraction (Figure 23). Cross-currents or occasional storm waves may assist this hooked formation. A good example is the sandspit in Walvis Bay, Namibia. Here the main body of the spit is curved but has additional, smaller hooks, or **recurves**. Longshore drift moves sediment northwards along the coast. However, the coastline is very irregular and there is a sudden change in the trend of the coastline. Consequently refraction occurs, causing the waves to bend around eastwards.

On the seaward side, the slope to deeper water is very steep. Within the curve of the spit, the water is shallow and a considerable area of mudflat and saltmarsh is exposed at low water. These saltmarshes are continuing to grow as mud is being trapped by the marsh vegetation.

Related features include bars. These are ridges that block off a bay or river mouth. There are many examples on the west coast of Antigua (Figure 24).

Tombolos are ridges that link the mainland to an island. Good examples include the Lumley area of Sierra Leone, and the Cape Verde Peninsula in Senegal. The Cape Peninsula in South Africa is a complex tombolo that has developed on a very large scale.

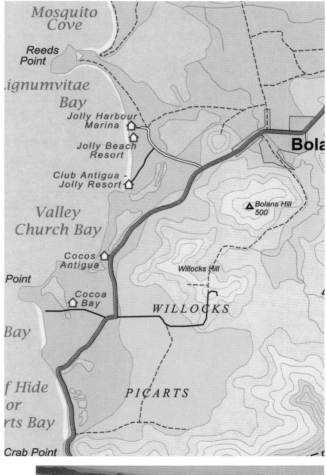

Figure 24 The west coast of Antigua

a A map extract
b Aerial view

Figure 23 Development of a spit

❶ Successive positions of the growing spit. The recurved end develops as a result of
❷ wave refraction and the occurrence of
❸ irregular winds from an alternative direction

River estuary

Original coastline

Headland

Saltmarsh

Position of fastest current

Longshore drift

Prevailing wind

Short-term change in wind and wave direction

Activities

1 a) Draw a labelled sketch of Figure 24b (page 79).
   b) Describe the wave conditions along this coastline.

2 Explain how a spit develops.

3 In what ways might vegetation help spits, bars and tombolos to develop?

4 Study Figure 24a. Name and give examples of at least two types of coastal deposit. For any one of these, describe its main characteristics and explain how it has been formed.

5 Study Figures 24a and b, and answer the following questions.

   a) What type of feature is found at i) Valley Church Bay and ii) Reeds Point?
   b) What is the difference between a cove and a bay?
   c) How are land-based processes affecting this area of coastline?
   d) In what ways has this area of coastline influenced human activities?

Sand dunes and saltmarshes
Succession on sand dunes

Sand dunes are one of the most dynamic environments in physical geography. Important changes take place in a very short space of time. Extensive sandy beaches are almost always backed by sand dunes because strong onshore winds can easily transport the sand which has dried out and is exposed at low water. The sand grains are trapped and deposited against any obstacle on land, to form dunes. Dunes can be blown inland and can therefore threaten coastal farmland and even villages. The interaction of winds and vegetation helps form sand dunes.

On the beach conditions are very windy, dry (much water just soaks into the sand) and salty. Few plants can survive these extreme conditions. Two plants that can are sea couch and marram grass (Figure 26). These are adapted to tolerate water with a high salt content and high wind speeds, and

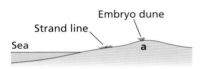

As the tide goes out, the sand dries out and is blown up the beach. A small embryo dune forms in the shelter behind the strand line.

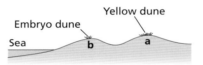

Sea couch grass colonises and helps bind the sand. Once the dune grows to over 1m high, marram grass replaces the sea couch.

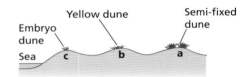

Once the yellow dune is over 10m high, less sand builds up behind it and marram grass dies to form a thin humus layer. As the original dune a has developed, new embryo and yellow dunes have formed.

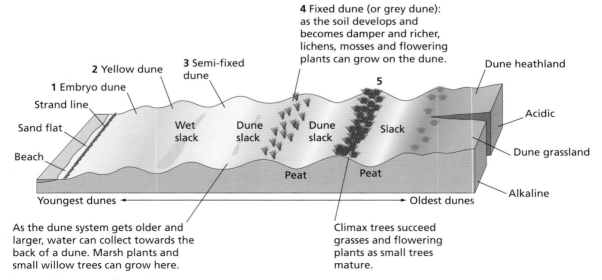

As the dune system gets older and larger, water can collect towards the back of a dune. Marsh plants and small willow trees can grow here.

Climax trees succeed grasses and flowering plants as small trees mature.

Figure 25 Development of sand dunes

they can survive burial by sand. In fact, marram needs to be buried by fresh sand in order to send out fresh shoots.

Once marram and sea couch have established in the beach, they reduce the wind speed and this helps trap fresh sand. As the sand builds up, the marram and couch send out new shoots, trapping more sand and building up a dune. Increasingly, the presence of marram and couch in the sand dune adds organic matter and moisture to the sand dune and allows other plants to grow, such as heather.

The growth of new plants is called **succession**. Plants such as heather cannot tolerate the dry, windy, salty conditions of the beach but can survive in the less windy, moister, less salty dunes. They in turn alter the environment so that other species can invade and develop. On a sand dune over a distance of just a few hundred metres there may be as many as four or five different types of ecosystems.

Saltmarshes

Saltmarshes are typically found in three locations: on low-energy coastlines, behind spits and barrier islands, and in estuaries and harbours. Silt accumulates in these situations and, on reaching sea level, forms mudbanks. With the appearance of vegetation a saltmarsh is formed. The mudbanks are often intersected by creeks.

Once the saltmarsh flat is formed, the first plants, such as green algae, colonise the mudflat (Figure 27). The algae trap sediment from the sea and provide ideal conditions for the seeds of the salt-tolerant marsh samphire and eel grass which then colonise the marsh. Halophytic (salt-tolerant) plants have adapted to the unstable, rapidly changing conditions. These plants increase the rate of deposition by slowing down the water as it passes over the vegetation. This is known as **bio-construction**. Gradually, the clumps of vegetation become larger and the flow of tidal waters is restricted to specific channels, or creeks (Figure 28, page 82). The slightly increased height of the surface around plants leads to more favourable conditions. Here plants are covered by seawater for shorter periods of time and this encourages other plants to colonise, such as sea aster, sea poa and sea blite. These are even more efficient at trapping sediment and the height of the saltmarsh increases. New plants colonise as the marsh grows, including sea lavender, sea pink and sea purslane. As the height increases, tidal flooding of the marsh becomes less frequent and the rate of growth slows down. Sea rush and black saltwort become the most common type of plants. It takes about 200 years to progress from the marsh samphire stage to the sea rush stage.

Figure 26 Sand dune vegetation

Figure 27 Early saltmarsh colonisers

Figure 28 Saltmarsh vegetation rapidly increases the deposition of mud

Activities

Study Figure 25 (page 80).

1 How do conditions differ between **a)** the shoreline and **b)** the saltmarsh?

2 Suggest why deposition occurs **a)** on the sand dunes and **b)** in the saltmarsh.

3 Where is the source of material for **a)** the sand dune and **b)** the saltmarsh?

4 How does vegetation help to build up the sand dunes and the saltmarsh?

5 Suggest how human activities might affect the sand dunes and/or the saltmarsh.

Coral reefs

Coral reefs are calcium carbonate structures, made up of reef-building stony corals. Coral is limited to the depth that light can reach and so reefs occur in shallow water, ranging to depths of 60 m. This dependence on light also means reefs are only found where the surrounding waters contain relatively small amounts of suspended material. Reef-building corals live only in tropical seas, where temperature, salinity and clear water allow them to develop.

There are several different types of coral reef (Figure 29).

- **Fringing reefs** are those that fringe the coast of a landmass (Figures 30 and 31). Many fringing reefs grow along shores that are protected by barrier reefs and are thus characterised by organisms that are best adapted to low wave-energy conditions.

- **Barrier reefs** occur at greater distances from the shore than fringing reefs and are commonly separated from it by a wide deep lagoon. Barrier reefs tend to be broader, older and more continuous than fringing reefs. The Beqa barrier reef of Fiji stretches unbroken for more than 37 km; the one off Mayotte in the Indian Ocean for around 18 km. The largest barrier-reef system in the world is the Great Barrier Reef, which extends 1600 km along the east Australian coast, usually tens of kilometres offshore. Another long barrier reef is located in the Caribbean off the coast of Belize between Mexico and Guatemala.

- **Atoll reefs** rise from submerged volcanic foundations. Atoll reefs are essentially indistinguishable in form and species composition from barrier reefs except that they are confined to

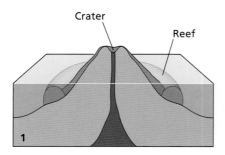

Rocky volcanic islet encircled by fringing coral reef

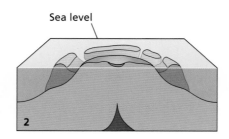

Reef enlarges as land sinks (or sea rises)

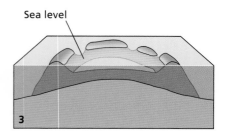

Circular coral reef or atoll (with further change in level)

Figure 29 The formation of coral reefs

the flanks of submerged oceanic islands, whereas barrier reefs may also flank continents. There are more than 300 atolls in the Indian/Pacific Oceans but only 10 in the western Atlantic.

Coral reefs are often described as the 'rainforests of the sea' on account of their rich biodiversity. Some coral is believed to be 2 million years old, although most is less than 10,000 years old. Coral reefs together support nearly a million species of plants and animals, and about 25 per cent of the world's sea fish breed, grow, spawn and evade predators in coral reefs. Some of the world's best coral reefs include Australia's Great Barrier Reef, many of the reefs around the Philippines and Indonesia, Tanzania and the Comoros, and the Lesser Antilles in the Caribbean.

Coral reefs today face many pressures. The fishing industry now uses dynamite to flush out fish, and cyanide solution to catch live fish. Destruction takes many forms – collection of specimens, trampling, berthing of boats, oil spills, mining for building and the cement industry. In addition, indirect pressures include sedimentation from rivers and waste disposal from urban areas. Coastal development, especially for tourism, is taking its toll too. Dust storms from the Sahara have introduced bacteria into Caribbean coral, while global warming may cause coral bleaching. **Bleaching** occurs when high temperatures expel the algae in coral, removing their colour – hence the coral appears bleached. Many areas of coral in the Indian Ocean were destroyed by the 2004 tsunami.

Coral reefs are of major biological and economic importance. Countries such as Barbados, the Seychelles and the Maldives rely on tourism. Florida's reefs attract tourism worth US$1.6 billion annually. The global value of coral reefs in terms of fisheries, tourism and coastal protection is estimated to be US$375 billion!

Figure 30 A fringing reef on the south coast of Antigua

Figure 31 A fringing reef off the west coast of Antigua

Activities

1 Describe the conditions needed for coral to grow.

2 What is the difference between a fringing reef and a barrier reef?

3 How are atolls formed?

4 Why are coral reefs so valuable?

5 What are the main threats to coral reefs?

Weather, climate and natural vegetation

The weather station

A weather station is a place where the elements of weather – temperature, humidity, pressure, wind direction and velocity, cloud cover and sunshine – are measured and recorded as accurately as possible. The weather station comprises an open piece of land and it contains the following instruments: thermometers (kept in a Stevenson screen, Figure 1), a rain gauge, a wind vane, an anemometer, a sunshine recorder and barometers.

The Stevenson screen is a wooden box standing on four legs at a height of about 120 cm. The screen is built so that the shade temperature of the air can be measured. The sides of the box have slats (louvers) to allow free entry of air, and the roof is made of double boarding to prevent the sun's heat from reaching the inside of the screen. Insulation is further improved by painting the outside of the screen white so as to reflect much of the sun's energy. The screen is usually placed on a grass-covered surface thereby reducing the radiation of heat from the ground.

Instruments kept inside the Stevenson screen include a maximum and minimum thermometer and a wet- and dry-bulb thermometer (also called a hygrometer) (Figure 2 and Figure 5, page 87).

Instruments placed outside the Stevenson screen include a rain gauge, a wind vane to determine wind direction, and an anemometer to assess wind speed. The rain gauge is kept well clear of trees and buildings so that only raindrops falling directly

Figure 1 A Stevenson screen

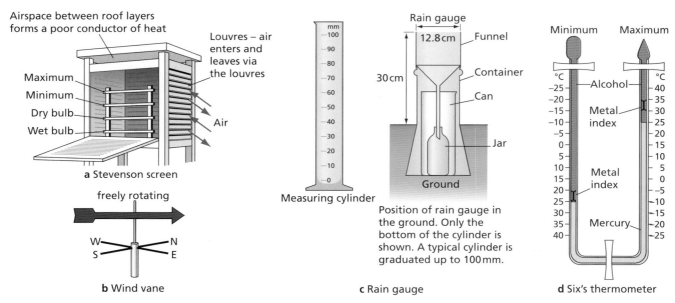

a Stevenson screen

Airspace between roof layers forms a poor conductor of heat

Louvres – air enters and leaves via the louvres

Maximum
Minimum
Dry bulb
Wet bulb

Air

b Wind vane

freely rotating

W — N
S — E

Measuring cylinder

mm
—100
—90
—80
—70
—60
—50
—40
—30
—20
—10
—0

c Rain gauge

Rain gauge

12.8 cm — Funnel

30 cm — Container

Can

Jar

Ground

Position of rain gauge in the ground. Only the bottom of the cylinder is shown. A typical cylinder is graduated up to 100 mm.

d Six's thermometer

Minimum   Maximum

°C
—25
—20
—15
—10
—5
0
5
10
15
20
25
30
35
40

Alcohol

Metal index

Metal index

Mercury

°C
40
35
30
25
20
15
10
5
0
—5
—10
—15
—20
—25

Figure 2 Equipment in a weather station

from the sky enter the funnel of the gauge. The wind vane and anemometer are placed well away from any buildings or trees that might interfere with the free movement of air. Buildings can channel air through narrow passages between two buildings, or decrease the flow of air by blocking its path. Trees have a similar effect.

A rain gauge is used to measure rainfall. It consists of a cylindrical copper container, in which there is a collecting can containing a glass or plastic jar, and a funnel that fits on to the top of the container. The gauge is sunk into the ground so that the top of the funnel is about 30 cm above ground level.

Rain falling over the funnel collects in the jar. This is emptied, usually every 24 hours, and measured in a tapered glass measure, graduated in millimetres. The tapered end of the jar enables very small amounts of rain to be measured accurately.

The gauge is placed in an open space so that no runoff from trees or buildings or other objects can get into the funnel. Also, the outer case is sunk into the ground to prevent the sun's heat from evaporating any of the rain collected in the glass jar. It projects about 30 cm above ground level to prevent any rain from splashing up from the ground into the funnel. The rainfall recorded for a place, either for a day or for a week or a month or a year, can be shown on a map. This is done by using lines called **isohyets**. An isohyet is a line on a map that joins areas of equal rainfall.

Measurement of air pressure

Because air has weight it exerts a pressure on the Earth's surface. At sea level the pressure is about $1.03 \, kg/cm^2$. Pressure varies with temperature and altitude, and the instrument that measures pressure is called a **barometer**. Air pressure is usually measured in units called **millibars**.

A **mercury barometer** consists of a hollow tube from which the air is extracted before the open end is placed in a bath of mercury. Mercury is forced up the tube by the pressure of the atmosphere on the mercury in the bath. When the pressure of the mercury in the tube balances the pressure of the air on the exposed mercury, the mercury in the tube stops rising. The height of the column of mercury changes as air pressure changes – that is, it rises when air pressure increases and falls when air pressure decreases.

An **aneroid barometer** is an instrument that consists of a small metal cylinder which is a vacuum chamber. A strong metal spring prevents the chamber from collapsing. The spring contracts and expands with changes in atmospheric pressure. These changes are magnified by a series of levers and they are conveyed to a pointer which moves across a calibrated scale. A **barograph** is an aneroid barometer which continuously records for one week. Changes in pressure are recorded by a flexible arm which traces an ink line on a rotating paper-covered drum. The paper is divided by vertical lines at two-hour intervals.

The atmospheric pressure is recorded at numerous weather stations for a region and these are plotted on a map of the region. Before this is done, the pressures are 'reduced' to sea level – that is, they are adjusted to what they would be if the stations were at sea level. The pressures are plotted on a map. Lines are then drawn through points whose pressure is the same. These lines are called **isobars**.

Measurement of wind direction and velocity

A **wind vane** is used to indicate wind direction. It consists of a horizontal rotating arm pivoted on a vertical shaft. The rotating arm has a tail at one end and a pointer at the other. When the wind blows, the arm swings until the pointer faces the wind. The directions north, east, south and west are marked on arms that are rigidly fixed to the shaft.

The speed of the wind is measured by an **anemometer** (Figure 3), which consists of three or four metal cups fixed to metal arms that rotate freely on a vertical shaft. When there is a wind, the cups rotate. The stronger the wind, the faster the rotation. The number of rotations are recorded on a meter to give the speed of the wind in km/hour.

Figure 3 An anemometer

Winds are shown by arrows on a weather map. The shaft of an arrow shows wind direction and the feathers on the shaft indicate wind velocity. The tip of the arrow (at the opposite end from the feathers) points to the direction in which the wind is blowing.

Wind direction for a specific place can be shown on a **wind rose**. It is made up of a circle from which rectangles radiate (see Figure 9, page 88). The directions of the rectangles represent the points of the compass. The lengths of the rectangles are determined by the number of days the wind blows from that direction. The number of days when there is no wind is recorded in the centre of the rose.

Measurement of temperature

Variations in temperature represent responses to differences in insolation. Temperature is measured using a **thermometer**. A continuous temperature reading is given by a **thermograph**.

Maximum thermometer

When the temperature rises, the mercury expands and pushes the index along the tube. When the temperature falls, the mercury contracts and the index remains behind. The maximum temperature is obtained by reading the scale at the end of the index which was in contact with the mercury. The index is then drawn back to the mercury by a magnet.

Minimum thermometer

When the temperature falls, the alcohol contracts and its meniscus pulls the index along the tube. When the temperature rises, the alcohol expands. The daily readings of the maximum and minimum thermometers are used to work out the average or mean temperature for one day (called the **mean daily temperature**) and the temperature range for one day (called the **daily** or **diurnal temperature range**).

To find the mean daily temperature the maximum and minimum temperatures for one day are added together and then halved; for example, maximum temperature 35 °C, minimum temperature 25 °C, mean daily temperature is 30 °C. The sum of the daily mean temperatures for one month divided by the number of days for that month gives the mean monthly temperature. The sum of the mean monthly temperatures divided by 12 gives the mean annual temperature.

The daily or diurnal temperature range is found by subtracting the minimum temperature from the maximum temperature for any one day; for example maximum temperature 35 °C, minimum temperature 25 °C, daily or diurnal temperature range 10 °C.

The highest mean monthly temperature minus the lowest mean monthly temperature gives the mean annual temperature range. For example, Lagos has a mean maximum temperature of 27.5 °C (March), and a mean minimum temperature of 24.5 °C (August). Its mean annual temperature range is therefore 3 °C.

A Six's thermometer (see Figure 2d, page 84) can be used to measure maximum and minimum temperatures at the same time. When the temperature rises, the alcohol in the left-hand limb expands and pushes the mercury down the left-hand limb and up the right-hand limb. The maximum temperature is read from the scale on the right-hand limb. When the temperature falls, the alcohol in the left-hand limb contracts and some of the alcohol vapour in the conical bulb liquefies. This causes the mercury to flow in the reverse direction. A metal index in each limb marks the temperature reached.

How is relative humidity measured?

Wet- and dry-bulb thermometers are used to measure relative humidity. The dry-bulb is a glass thermometer that records the actual air temperature. The wet-bulb is a similar thermometer, but with the bulb enclosed in a muslin bag which dips in to a bottle of water. This thermometer measures the wet-bulb temperature which, unless the relative humidity is close to 100%, is generally lower than the dry-bulb temperature. Absolute humidity refers to the amount of moisture in the atmosphere, for example, 10 g/m^{-3}. Relative humidity expresses how much is in the atmosphere compared with the maximum that air of a given temperature could hold. For example, air of 20 °C with 8.5 g of moisture has a relative humidity of $8.5/17 \times 100 = 50\%$.

When the wet- and dry-bulb readings are the same, the relative humidity is 100 %. As the difference between the wet- and dry-bulb temperatures increases, the relative humidity is lower.

Dry bulb temperature	1.0	2.0	3.0	4.0
20	91	81	73	64
15	89	78	68	58
10	87	74	62	50

Figure 4 Difference between wet- and dry-bulb readings (the figure in the table is the relative humidity)

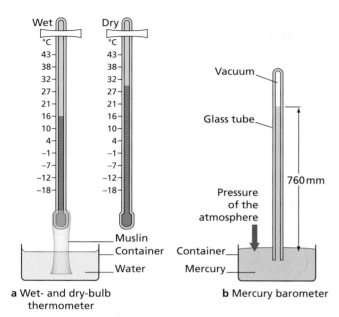

a Wet- and dry-bulb thermometer

b Mercury barometer

Figure 5 More measuring instruments

Activities

1 Describe and explain the main characteristics of a Stevenson screen.

2 What information does a Six's thermometer show?

3 Why are weather readings taken at the same time each day?

4 Where is the best place to locate a rain gauge? Briefly explain why.

5 How are wind speed and wind direction measured?

Recording the weather

Clouds

The ten main types of clouds can be separated into three broad categories according to the height of their base above the ground: high clouds, medium clouds and low clouds (Figure 6).

High clouds are usually composed solely of ice crystals and have a base between 5500 and 14,000 m:

- cirrus – white filaments
- cirrocumulus – small rippled elements
- cirrostratus – transparent sheet, often with a halo.

Medium clouds are usually composed of water droplets or a mixture of water droplets and ice crystals, and have a base between 2000 and 7000 m:

- altocumulus – layered, rippled elements, generally white with some shading
- altostratus – thin layer, grey, allows sun to appear as if through ground glass
- nimbostratus – thick layer, low base, dark, rain or snow may fall from it.

Low clouds are usually composed of water droplets, though cumulonimbus clouds include ice crystals, and have a base below 2000 m:

- stratocumulus – layered, series of rounded rolls, generally white with some shading
- stratus – layered, uniform base, grey
- cumulus – individual cells, vertical rolls or towers, flat base
- cumulonimbus – large cauliflower-shaped towers, often 'anvil tops', sometimes giving thunderstorms or showers of rain or snow.

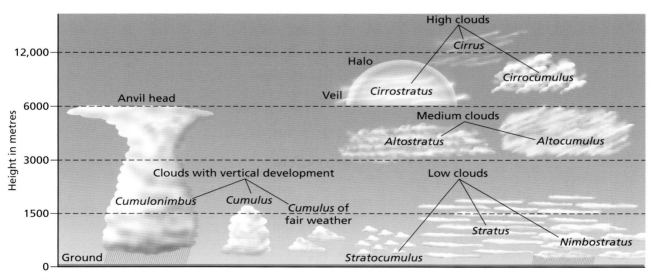

Figure 6 Cloud types

Date	Day	Temperature		Rainfall mm	Wind direction	Wind speed km/hour	Air pressure mb
		Max. °C	Min °C				
1 August	W	14.2	9.7	4.0	N	22	1006
2	Th	13.4	11.5	0	N	37	1004
3	F	9.9	8.1	0	WNW	33	1011
4	S	11.5	7.2	0	WNW	31	1016
5	S	11.6	8.2	0	W	28	1019
6	M	12.7	9.5	20.2	W	20	1023
7	T	14.5	9.2	0	N	30	1019

Figure 7 Daily weather observations at Frankston, Victoria (Australia), 1 August 2007 – 7 August 2007

Date	Day	Temperature		Rainfall mm	Wind direction	Wind speed km/hour	Air pressure mb
		Max. °C	Min °C				
1 February	F	25.6	11.7	6.8	SSE	15	1020
2	S	25.7	16.9	0	NNW	9	1016
3	S	27.6	17.9	0	SE	9	1016
4	M	29.1	19.9	0	ENE	11	1013
5	T	23.2	19.7	0	SW	13	1012
6	W	23.1	19.2	0	SW	19	1004
7	Th	17.9	15.7	8.4	SW	19	1005

Figure 8 Daily weather observations at Frankston, Victoria (Australia), 1 February 2008 – 7 February 2008

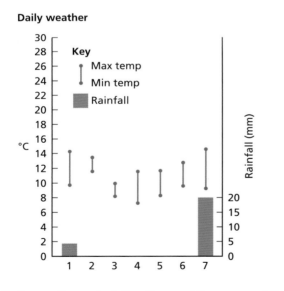

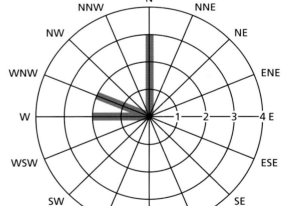

Figure 9 Daily weather, wind direction and frequency at Frankston, August 2007

Activities

The results recorded by a school in Victoria are shown in Figures 7 and 8. The data for the first week were plotted as in Figure 9.

1 Plot the data for February (Figure 8) using the same methods used to create Figure 9 (for August).

2 State the maximum and minimum temperatures of the seven-day period.

3 Work out the mean minimum temperature and the mean maximum temperature for the seven days.

4 How much rain fell during the seven days?

5 Compare the weather in February with that in August.

Climates associated with tropical rainforests and hot deserts

The main characteristics of the climate in tropical rainforests include:

- hot conditions – generally above 26 °C – throughout the year
- high levels of rainfall, often over 2000 mm
- a lack of seasons – the temperatures are high throughout the year
- the difference between daytime and night-time temperatures (the **diurnal range**) is, in fact, higher than the seasonal differences in temperature
- rainfall is mainly convectional and may fall on as many as 250 days each year
- cloud cover varies – in the morning it may be limited but by afternoon, towering cumulonimbus clouds mark the start of the convectional rains
- the presence of clouds tends to reduce the amount of heat that is lost at night – hence the diurnal range is less than in hot desert areas
- the humidity (moisture in the atmosphere) is high, and relative humidities of 100% are often reached in the late afternoon
- wind speeds within the forest are reduced by the large numbers of trees present.

The graph and data for Manaus in Brazil (Figure 10) show that the warmest months are September and October with a mean monthly temperature of 34 °C.

Rainfall in Manaus is high, nearly 2100 mm. There is a definite wet season between November and May. The temperatures are hot due to the equatorial location (the sun is always high in the sky) in all seasons. Warm temperatures encourage convectional uplift of air and hence heavy rain in most seasons.

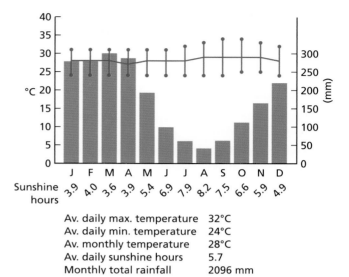

Sunshine hours: 3.9 4.0 3.6 3.9 5.4 6.9 7.9 8.2 7.5 6.6 5.9 4.9

Av. daily max. temperature	32°C
Av. daily min. temperature	24°C
Av. monthly temperature	28°C
Av. daily sunshine hours	5.7
Monthly total rainfall	2096 mm

Figure 10 Climate data for Manaus

In contrast, the main characteristics of hot desert climates include:

- very hot days and cold nights, caused by the lack of cloud cover
- low and irregular amounts of rainfall, which lack any seasonal pattern
- low levels of humidity for much of the year
- warm dry winds, sometimes causing sandstorms.

The data for Cairo (Figure 11) show that the warmest mean monthly temperatures are between June and August when the temperature reaches 35 °C. In contrast, the lowest mean monthly temperature is in January, reaching just 9 °C. Thus the temperature range is 26 °C. There is a seasonal pattern to temperature with the highest values in the summer and lowest readings in the winter. Rainfall figures are very low, just 27 mm. Temperatures are hot due to the position of the sun high overhead. Rainfall is limited because these

	J	F	M	A	M	J	J	A	S	O	N	D	Av/Total
Temperature													
Daily max. °C	19	21	24	28	32	35	35	35	33	30	26	21	28
Daily min. °C	9	9	12	14	18	20	22	22	20	18	14	10	16
Average monthly °C	14	15	18	21	25	28	29	28	26	24	20	16	22
Rainfall													
Monthly total mm	4	4	3	1	2	1	0	0	1	1	3	7	27
Sunshine hours	6.9	8.4	8.7	9.7	10.5	11.9	11.7	11.3	10.4	9.4	8.3	6.4	9.5

Figure 11 Climate data for Cairo

regions are areas of descending air – this prevents rain from forming.

Activities

Refer to Figures 10 and 11 (page 89).

1 Plot the data for Cairo (Figure 11) as presented in Figure 10 for Manaus.

2 In which months is the average temperature in Cairo higher than in Manaus?

3 How much rain falls in Manaus in April?

4 In which months is the minimum temperature in Cairo higher than in Manaus? How do you explain this answer?

5 Describe the variations in monthly sunshine levels in Manaus.

6 Suggest why there is a link between sunshine levels and rainfall.

7 What is the mean monthly temperature range in Cairo and Manaus in a) July and b) December?

■ Factors affecting climate

Many factors affect the temperature of a place. These include latitude, distance from the sea, the nature of nearby ocean currents, altitude, dominant winds, cloud cover, and aspect. Differences in pressure systems also affect whether it rains or is dry.

Latitude

On a global scale latitude is the most important factor determining temperature (Figure 12), in two ways. Firstly, at the equator the overhead sun is high in the sky, so high-intensity insolation is received; by contrast, at the poles the overhead sun is low in the sky, so the quantity of energy received is low. Secondly, the thickness of the atmosphere affects temperature. Near the poles the sun's rays have more atmosphere to pass through, due to the low angle of approach. So more energy is lost, scattered or reflected over polar regions than over equatorial areas, making temperatures lower over the poles.

Proximity to the sea

The **specific heat capacity** is the amount of heat needed to raise the temperature of a body by 1 °C. Land heats and cools more quickly than water. It takes five times as much heat to raise the temperature of water by 1 °C as it does to raise land temperatures.

Water also heats more slowly because:

- it is clear, hence the sun's rays penetrate to greater depth (distributing energy over a wider area)
- tides and currents cause the heat to be distributed further.

A greater volume of water is therefore heated for every unit of energy than land, so water takes longer to heat up.

Therefore distance from the sea has an important influence on temperature. Water takes up heat and emits it much more slowly than the land. In winter in mid-latitudes, sea air is much warmer than the land air, so onshore winds bring heat to the coastal lands. By contrast, during the summer coastal areas

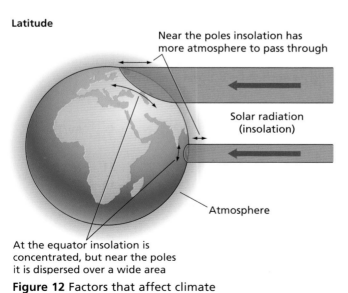

Figure 12 Factors that affect climate

remain much cooler than inland sites. Areas with a coastal influence are termed **maritime** or **oceanic** whereas inland areas are described as **continental**.

Ocean currents

The effect of ocean currents on temperatures depends upon whether the current is cold or warm. Warm currents from equatorial regions raise the temperatures of polar areas (with the aid of prevailing westerly winds). However, the effect is only noticeable in winter.

Altitude

In general, air temperature decreases with increasing altitude. This is because air under the greater pressure of lower altitudes is denser and therefore warmer. As altitude increases so the pressure on the air is reduced and the air becomes cooler. The normal decrease of temperature with height is on average 10 °C/km.

Winds

The effects of winds on temperature depends on the initial characteristics of the wind. In temperate latitudes **prevailing** (dominant) winds from the land lower the winter temperatures but raise the summer ones. This is because continental areas are very hot in summer but very cold in winter. Prevailing winds from the sea do the opposite: lower the summer temperatures and raise the winter ones.

Cloud cover

Cloud cover decreases the amount of insolation reaching the surface by reflecting some of it. Clouds also reduce the amount of insolation leaving the surface by absorbing the radiation. If there is no cloud then incoming short-wave radiation and outgoing long-wave radiation are at a maximum.

Pressure

In low pressure systems air is rising. It may rise high enough to cool, condense and form clouds and rain. This can happen in very warm areas, such as rainforests, at mountain barriers and at weather fronts, when warm air is forced over cold air. In contrast, where there is high pressure air is sinking, and rain formation is prevented. The world's great hot deserts are located where there is high pressure caused by sinking air.

Activities

1 How does latitude affect the amount of heat a place receives?

2 Why are equatorial areas not getting any hotter nor polar areas any colder?

3 What is meant by the term *specific heat capacity*?

4 Explain why temperature decreases with height.

5 Why do deserts experience a large difference in temperature between day and night but rainforest areas do not?

■ Tropical rainforests

Tropical evergreen rainforests are located in equatorial areas, largely between 10°N and 10°S. There are, nevertheless, some areas of rainforest that are found outside these areas but these tend to be more seasonal in nature. The main areas of rainforest include the Amazon rainforest in Brazil, the Congo rainforest in central Africa, and the Indonesian-Malaysian rainforests of south-east Asia. There are many small fragments of rainforest, such as those on the island of Madagascar and in the Caribbean. Everywhere tropical rainforests are under increasing threats from human activities such as farming and logging. The result is that rainforests are disappearing and those that remain are not only smaller, but broken up into fragments.

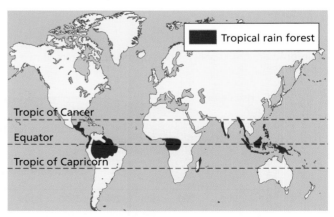

Figure 13 World distribution of tropical rainforests

Vegetation

The vegetation is evergreen, enabling photosynthesis to take place year round. This is possible because temperatures are high all year, and water is available throughout the year. The vegetation is layered, and the shape of the crowns varies with the layer, so that plants can receive light. Species at the top of the canopy receive most of the sunlight whereas species that are located near the forest floor are adapted to darker conditions, and generally have a darker pigment so as to photosynthesise in low light levels. There is a great variety in the number of species in a rainforest – this is known as **biodiversity**. A rainforest may contain as many as 300 different plant species in a single hectare. Typical rainforest species include figs, teak, mahogany and yellow woods.

Tropical vegetation has many adaptations (Figure 15). Many trees have leaves with drip-tips. These are designed to get rid of excess moisture. In contrast, other plants have saucer-shaped leaves in order to collect water. Pitcher plants have developed an unusual means of getting their nutrients. Rather than getting nutrients from the soil, they have become carnivorous and acquire their nutrients from insects and small frogs that get inside the pitcher. This is one way of coping with the very infertile soils of the rainforest. Other plants are very tall. To prevent them being blown over by the wind, some plants have developed buttress roots which project out from the main trunk above ground thereby giving the plant extra leverage in the wind.

Rainforests are the most productive land-based ecosystem. Ironically, the soils of tropical rainforests are quite infertile. This is because most of the nutrients in the rainforest are contained in the **biomass** (living matter). Rainforest soils are typically deep due to the large amount of weathering that has taken place, and they are often red in colour, due to the large amounts of iron present in the soil. Nevertheless, there are some areas where tropical soils may be more fertile – on floodplains the soils may be enriched by flooding and in volcanic areas by the weathering of fertile lava flows.

The nutrient cycle is one that is easily disrupted. Tropical rainforests have been described as 'deserts covered by trees'. Once the vegetation is removed, nutrients are quickly removed from the system creating infertile conditions, even deserts.

Figure 15 Rainforest plants – pitcher plant and drip tips

Rainforest are found only in areas with over 1700 mm of rain and temperatures of over 25 °C

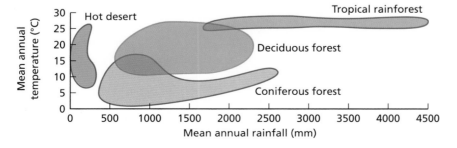

Figure 14 Conditions required for the growth of rainforests

The links between climate, soils and vegetation are very strong

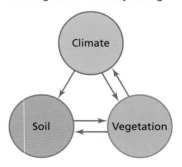

Hot deserts

The world's hot deserts are largely found in subtropical areas around 20°–30° north and south of the equator. The largest area of hot desert is the Sahara but there are other important deserts such as the Great Victorian Desert and Great Sandy Desert in Australia, the Kalahari and Namib Deserts in southern Africa, the Atacama Desert in South America, and the Arabian Desert. The Gobi Desert in Mongolia and China lies outside the tropics and therefore is not a hot desert.

The main characteristics of the climate that influence the vegetation are that it is hot throughout the year and there is low and unreliable rainfall (not more than 250 mm per year). The vegetation responds to this in a number of ways.

Plants

There are two main types of plants in deserts. Perennials (plants which grow over a number of years) may be succulent (they store lots of water), they are often small (to reduce water loss by evaporation and transpiration) and they may be woody. Annuals or ephemerals are plants that live for a short time but they may form a dense covering of vegetation immediately after rain.

Ephemerals evade drought. During the infrequent wet periods they develop rapidly producing a large number of flowers and fruits. These help produce seeds which remain dormant in the ground until the next rains.

Many plants are adapted to drought – these are called **xerophytes**. Water loss is minimised in a number of ways:

- leaf hairs reduce wind speed and therefore reduce transpiration
- thick waxy cuticles and the rolling-up or shedding of leaves at the start of the dry season decreases water loss
- some plants have the bulk of their biomass (living material) below the ground surface
- others have very deep roots to reach the water table
- the wood in woody species prevents the collapse of the plant even when the plant is wilting, for example during a drought.

Vegetation from desert margins is often referred to as **scrub**. Tropical scrub on the margins of hot deserts includes acacia, cactus, succulents, tuberous-rooted plants and herbaceous plants which only grow with rain. Special types are *mulga* in Australia (dense acacia thickets), *spinifex* in Australia ('porcupine grass'), and *chanaral* in Chile (spiny shrubs). Soils in desert areas are very infertile. As a result of the low rainfall there is little organic or moisture content in the soil. There is a lack of chemical weathering (largely due to the lack of moisture) so soils contain few nutrients.

Figure 16 Plants adapted to hot desert environments

Animals

Animals are adapted to living in the desert in a number of ways. These include:

- nocturnal (night-time) activity to avoid the heat of the day
- panting and/or large ears to reduce body heat
- burrowing by day
- secreting highly concentrated uric acid, thereby reducing water loss
- seasonal migration
- long-term aestivation (dormancy) which ends only when triggered by moisture and temperature conditions.

Activities

1 Why is it difficult to live in a hot desert?

2 Study Figure 14 on page 92 showing the conditions required for the growth of hot deserts.

   a) What is the maximum rainfall in a hot desert, as suggested by Figure 14?
   b) What is the range of mean annual temperatures in hot deserts?
   c) Suggest how a hot desert with a mean annual temperature of 30 °C and an annual rainfall of 250 mm might differ from one in which the mean annual rainfall is 250 mm and the mean annual temperature is 20 °C.

3 How have plants adapted to survive in the desert?

4 How do animals survive in the desert?

5 Search the internet to find out how camels are adapted to living in hot deserts.

Inter-relationships of physical and human geography

Volcanic eruptions

All natural environments provide opportunities and challenges for human activities. Some of the challenges can be described as 'natural hazards'. A **natural hazard** is a natural event that causes damage to property and/or disruption to normal life and may cause loss of life. Natural hazards involve hydrological, atmospheric and geological events. Natural hazards are caused by the impact of natural events on the social and economic environment in which people live. Some groups of people are more vulnerable to natural hazards and have greater exposure to them.

Since the 1960s more people have been affected by natural hazards. Reasons for this include:

- a rapid increase in population, especially in developing countries
- increased levels of urbanisation, including more shanty towns which are often located in hazardous environments
- changing land use in rural areas which results in flash floods, soil erosion and landslides
- increased numbers of people living in poverty who lack the resources to cope with natural hazards
- changes in the natural environment causing increased frequency and intensity of storms, floods and droughts.

> A **hazard** refers to a potentially dangerous event or process. It becomes a **disaster** when it affects people and their property.
>
> **Risk** suggests that there is a possibility of loss of life or damage. **Risk assessment** is the study of the costs and benefits of living in a particular environment.

There are two very different ways of looking at people's vulnerability:

- One view is that people choose to live in hazardous environments because they understand the environment. In this situation people choose to live in an area because they feel the benefits outweigh the risks.
- Another view is that some people live in hazardous environments because they have very little choice over where they live, as they are too poor to move.

Figure 1 Tourists gather around the geyser at Geysir, Iceland – one of the benefits of tectonic activity

Figure 2 A church buried by a mudflow in Plymouth, Montserrat – one of the disadvantages of tectonic activity

Direct hazards	Indirect hazards	Socio-economic impacts	
Pyroclastic flows	Atmospheric ash fallout	Destruction of settlements	
Volcanic bombs (projectiles)	Landslides	Loss of life	
Lava flows	Tsunamis	Loss of farmland and forests	
Ash fallout	Acid rainfall	Destruction of infrastructure – roads, airstrips and port facilities	
Volcanic gases		Disruption of communications	
Lahars (mudflows)			
Earthquakes			

Figure 3 Hazards associated with volcanic activity

Activities

1 What is a natural hazard?

2 Suggest reasons why natural hazards appear to be increasing in frequency.

3 How may volcanic activity be a benefit to people? (See also pages 59–62.)

4 Describe the direct and indirect hazards associated with volcanic activity in the Caribbean.

5 What are the potential impacts of volcanic activity on people's lives and livelihoods?

■ Tectonic hazards in the Caribbean

The Caribbean plate is one of the smaller surface plates of the Earth. Earthquakes occur all around its periphery, and volcanoes erupt on its eastern and western sides. The Caribbean plate moves more slowly, at about 1–2 cm a year, while the North American plate moves westward at about 3–4 cm a year). Many earthquakes and tsunamis have occurred in the north-eastern Caribbean region, where the movement of plates is rapid and complicated. There are a number of hazards related to earthquakes (Figure 4) and the impact on the death rate has been significant (Figure 5).

There are 25 potentially active volcanoes in the Caribbean, all of them in the eastern Caribbean. There have been 17 eruptions in recorded history: Mt Pelée (1902) accounted for most deaths and the Soufrière Hills volcano has been active since 1995. Kick 'em Jenny is an active submarine volcano, north of Grenada. All of the volcanoes are associated with subduction zones.

St John's Cathedral, Antigua rebuilt following the devastating earthquake of 1843

Primary hazard	Secondary hazard	Impacts	
Ground shaking	Ground failure and soil liquefaction	Total or partial destruction of building structures	
Surface faulting	Landslides and rockfalls	Interruption of water supplies	
	Debris flows and mud flows	Breakage of sewage disposal systems	
	Tsunamis	Loss of public utilities such as electricity and gas	
		Floods from collapsed dams	
		Release of hazardous material	
		Fires	
		Spread of chronic illness	

Figure 4 Earthquake hazards and impacts

Date	Location	Deaths	Magnitude
1902	Guatemala	2000	7.5
1907	Jamaica	1600	6.5
1918	Puerto Rico	116	7.5
1931	Nicaragua	2400	5.6
1946	Dominican Republic	100	8.0
1972	Nicaragua	5000	6.2
1976	Guatemala	23,000	7.5
1985	Mexico	9500	8.0
1986	El Salvador	1000	5.5
2001	El Salvador	844	7.7
2001	El Salvador	315	6.6

Figure 5 Major earthquakes in the Caribbean

Activities

1 Study Figure 6.

   a) What are the two plates responsible for tectonic activity in the Soufrière Hills?
   b) Which two plates are likely to have caused the earthquake that affected Mexico City in 1985?
   c) Describe what happens when the North American plate meets the Caribbean plate.

2 **a)** Describe the main hazards associated with earthquakes.
   b) Briefly explain any three of the impacts of earthquakes.

3 **a)** Choose a suitable diagrammatic method to show the relationship between the magnitude of an earthquake and the loss of life, as shown in Figure 5.
   b) Describe the relationship shown in your diagram.
   c) Suggest reasons for the relationship (or lack of) in your answer to 3a.

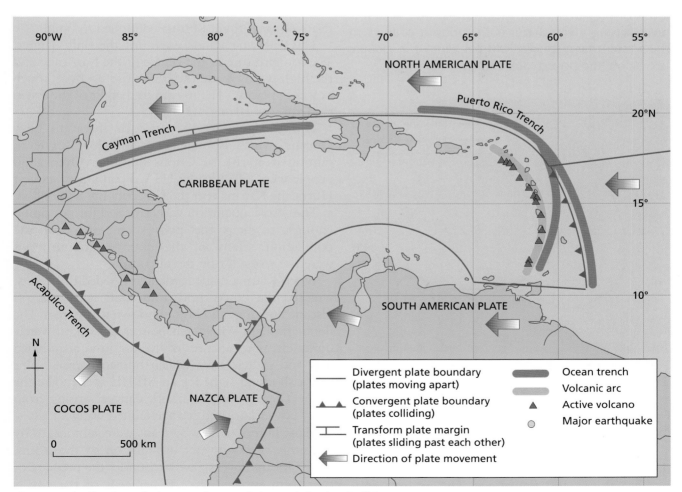

Figure 6 Distribution of plates and tectonic hazards in the Caribbean

Tropical storms

A hurricane is one of the most dangerous natural hazards to people and the environment. Damage is caused by high winds, floods and storm surges. Hurricanes are also known as **tropical cyclones** in Asia.

Hurricane Jeanne, September 2004

Hurricane Jeanne (Figure 7) was the sixth major hurricane of 2004. It was also the deadliest, with over 3000 deaths in Haiti, and five in the USA. Jeanne showed that the strongest natural hazard events do not necessarily cause the greatest damage. Population density, building construction and hazard planning measures also helped to determine the outcome of the hazard. Haiti has very poor levels of preparedness for natural hazards, which is unsurprising given that it is the poorest country in the western hemisphere – 80 per cent of its people live below the poverty line.

Figure 7 The tracks of Hurricane Jeanne and Hurricane Katrina

The scale of the disaster was blamed on deforestation, which left communities vulnerable to flash floods and mudslides. The mountainous country which was once heavily forested has less than 2 per cent tree cover. This has led to severe soil erosion which allows water to rush off the steep slopes. Most trees have been cut down to make charcoal for cooking. A recent UN environmental report described Haiti as 'one of the most degraded countries in the world'.

Hurricane Katrina

Hurricane Katrina (Figure 7) was the USA's worst natural disaster in living memory. The storm hit land near New Orleans on 29 August 2005 at a speed of some 225 km/h. Katrina was a category 4 hurricane, but what set it apart from other hurricanes was the way it lingered rather than passing through. Unlike most hurricanes of this intensity, which are over in a relatively brief time, Katrina continued to build.

The battering winds were not the only danger. The low pressure at the centre of the hurricane and the high winds made the ocean rise up by as much as 9 m in places. (The maximum height of the Asian tsunami was 10 m.)

The hurricane was a particular threat to New Orleans, which is built on land below sea level, putting it at risk of serious flooding. Over 1830 people were killed in the USA. Economists suggest Hurricane Katrina cost the US economy $80 billion.

The rescue operation was criticised for not doing enough to help the poorest members of the population. Many of the poor neighbourhoods were the worst hit by the hurricane.

When Katrina made landfall, it flooded the streets, wrecked the power grid, tore roofs and walls off historic buildings and brought down many trees. The floods brought with them poisonous snakes, water-borne diseases, carcasses of livestock and abandoned pets and grotesquely swollen human corpses. This was a shocking sight for an MEDC society like the USA. There were also health dangers arising from fallen power lines and sewage-tainted water.

Many homes in New Orleans were submerged by the surge of floodwater brought on by the storm. New Orleans' levées gave way under the pressure of the storm surge. Over 70 per cent of the city is below sea level and 80 per cent of it went under water, with some sections as deep as 6 m. One of the areas worst affected was the Ninth Ward, a poor district to the east of the city's famous French Quarter. The floodwaters in New Orleans were ten times more toxic than is considered safe.

- Median household income in the most devastated neighbourhood was $32,000, or $10,000 less than the national average.
- 20% of households in the disaster area had no car, compared with 10 per cent nationwide.
- Nearly 25% of those living in the hardest-hit areas were below the poverty line – about double the national average.
- About 60% of the 700,000 people in the 36 neighbourhoods affected (in the states of Louisiana, Mississippi and Alabama) were from an ethnic minority. Nationwide, about one in three US citizens is from a racial minority.

Figure 8 Impact and social status of damage caused by Katrina

Activities

1 Describe the main impacts of Hurricane Katrina.

2 Why were its impacts so great?

3 Describe two similarities and two differences in the impacts of Hurricanes Jeanne and Katrina.

■ Floods

Case Study

Bangladesh floods

Eighty per cent of the total area of Bangladesh consists of alluvial floodplains, mainly of the three large rivers (Ganges, Brahmaputra and the Meghna) which rise outside the borders of the country. The relief of the floodplains is generally less than 1 m. Flooding is a natural occurrence, regularly affecting 30 per cent of territory following monsoon rainfall and tropical storm surges (river

floods start in March with snowmelt and continue in June–July from monsoon rain). Coastal flooding occurs because of the high tidal range of the Bay of Bengal (5.6 m); and an average of about 16 cyclones a year. The flood-prone area is occupied by 80 per cent of the population who are rural and live on raised earth mounds, with the infrastructure, railways and roads on embankments.

Causes of major floods

The 1998 floods in Bangladesh were the worst on record. The reasons for this are complex, but all the following factors played a role:

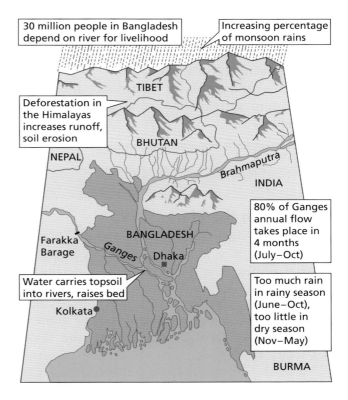

Impacts of the 1998 floods
■ Floods covered two-thirds of the country.
■ Nearly 5000 deaths.
■ 30 million people displaced.
■ 11,000 km of roads damaged.
■ 14,000 schools destroyed.
■ 660,000 ha of crops damaged.
■ Estimated economic cost U$5 billion.

Figure 9 Flooding in Bangladesh

- The nature of monsoon rain – this starts abruptly and is prolonged and very heavy. The highest rainfall totals in the world are experienced in the drainage basins of the Brahmaputra and the Meghna.
- The timing of the flood peaks of the three rivers – normally the flood peaks of the rivers do not coincide. In 1998, however, they peaked together.
- Build-up of silt in river beds – this has been due to heavy soil erosion following the development of agriculture into more marginal lands. River capacity was reduced, leaving them unable to cope with the increase in discharge. Deforestation in the upstream areas might have contributed to the build-up of silt in the river beds and more overland flow, but evidence is sparse and unreliable.
- Storm surge – this happened following a cyclone, which raised sea levels and prevented floodwaters from flowing into the sea.
- Human occupation of marginal lands – as population growth and rural impoverishment continued.

Case Study

Flooding on the Rhine, 1995

The Rhine is the longest river in Europe. It flows for 1320 km. Not only is the river an important physical feature, the drainage basin also contains 40 million people, and a large amount of industry. As a result of its importance, the Rhine has been heavily protected and engineered – but some of this work may have contributed to the floods of January 1995.

Causes

Natural causes included:

- heavy rain – Switzerland had over three times its average rainfall in January
- saturated soils – there was nowhere for the rain to soak away
- mild temperatures – this melted snow in the Alps.

There were also human causes:

- much of the Rhine's floodplain has been built upon – the impermeable surface increases the amount of rain now reaching the river, and the speed with which it does so

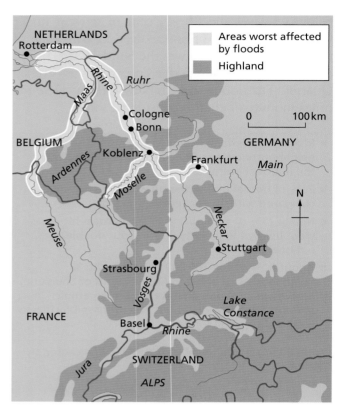

The 1995 floods

- 27 people were killed.
- Over 250,000 people were evacuated from their homes in the Netherlands.
- Flood damage in Germany alone was valued at over £640 million.

Figure 10 Flooding on the Rhine

- intensive farming compacts the soil and increases overland runoff
- vegetation clearance reduces interception
- channel straightening speeds up the flow of water downstream
- dykes (embankments) create faster and deeper flows.

Solutions to the flood problem

Short-term solutions include:

- evacuation of people and livestock
- sandbags placed across doors
- removal of furniture upstairs
- clearance of underground car parks and subways.

Long-term measures include:

- the development of an early warning system

- dykes to increase the volume of water the river can hold – the Dutch flood protection scheme has cost over £1 billion to build 600 km of dykes since 1995
- relief channels and basins to divert some of the water during the peak of the flood – but this requires co-operation between a number of countries in the upper course to prevent flooding in the lower course of the river
- artificial floodplains – called forelands in the Netherlands – located within the winter dykes; these areas are allowed to flood and can be used for grazing and recreation
- limited residential and industrial development in floodplain areas.

Activities

1 In what ways were the causes of the floods in Bangladesh and Europe a) similar and b) different?

2 For one of the floods, describe and suggest reasons for the impacts the flood had.

3 Suggest ways in which it might be possible to protect Bangladesh from flooding.

Drought

A large proportion of the world's surface experiences dry conditions. Semi-arid areas are commonly defined as having a rainfall of less than 500 mm per annum, while arid areas have less than 250 mm and extremely arid areas less than 125 mm per annum. In addition to low rainfall, dry areas have **variable rainfall**. As rainfall total decreases, variability increases. For example, areas with a rainfall of 500 mm have an annual variability of about 33 per cent. This means that in such areas rainfall could range from between 330 mm and 670 mm. This variability has important consequences for vegetation cover, farming and the risk of flooding.

Varieties and locations of dry areas

Arid conditions are caused by a number of factors:

- The main cause is the global atmospheric circulation. Dry, descending air associated with

the **subtropical high-pressure** belt is the main cause of aridity around 20°–30°N.
- In addition, distance from sea, **continentality**, limits the amount of water carried across by winds.
- In other areas, such as the Atacama and Namib Deserts, **cold offshore currents** limit the amount of condensation into the overlying air.
- Others are caused by intense **rainshadow effects**, as air passes over mountains. This is certainly true of the Patagonian Desert.
- A final cause, or range of causes, are human activities. Many of these have given rise to the spread of desert conditions into areas previously fit for agriculture. This is known as **desertification**, and is an increasing problem.

What is drought?

Drought is an extended period of dry weather leading to conditions of extreme dryness (Figure 11). Absolute drought is a period of at least 15 consecutive days with less than 0.2 mm of rainfall. Partial drought is a period of at least 29 consecutive days during which the average daily rainfall does not exceed 0.2 mm.

Figure 11 A dried-up river bed during a drought

Case Study

Europe's drought of 2003

Estimates for the death toll from the French heatwave in 2003 were as high as 30,000. Harvests were down between 30 per cent and 50 per cent on 2002. France's electricity grid was also affected as demand for electricity soared as the population turned up air conditioning and fridges. However, nuclear power stations, which generate around 75 per cent of France's electricity, were operating at a much reduced capacity because there was less water available for cooling.

Portugal declared a state of emergency after the worst forest fires for 30 years. Temperatures reached 43 °C in Lisbon in August 2003: 15 °C hotter than the average for the month. Over 1300 deaths occurred in the first half of August, and up to 35,000 hectares of forest, farmland and scrub was burned. Some fires were, in fact, deliberately started by arsonists seeking insurance or compensation money and more than 70 people were arrested. The prolonged heatwave left some countries facing their worst harvests since the end of the Second World War. Some countries that usually export food were forced to import it for the first time in decades. Across the EU, wheat production was down 10 million tonnes, about 10 per cent.

Case Study

Drought in Africa

In 2003 parts of southern Ethiopia were experiencing the longest drought anyone had known. The world's largest emergency food aid programme was in operation, but it proved inadequate. Because of a sixth poor rainy season in three years, 20 million people needed help. This was now worse than the 1984 famine, when only 10 million people needed food.

Ethiopia	20 million
Zimbabwe	7 million
Malawi	3.2 million
Sudan	2.9 million
Zambia	2.7 million
Angola	1.9 million
Eritrea	1 million
Plus around 7.3 million across Swaziland, Congo, Uganda, Congo Brazzaville, Lesotho and Mozambique	

Figure 13 Africa's 'at risk' population in 2003

Activities

1 Describe some of the hazards associated with drought.

2 Suggest reasons why the impact of drought may be increasing.

3 Suggest how the impact of the drought in Africa was different from the impact of drought in Europe.

Figure 12 A parched cricket ground, Oxford 2003

Opportunities and hazards in rainforests

About 200 million people live in areas that are or were covered by tropical rainforests. Tropical rainforests offer many advantages for human activities such as farming, hydro-electric power (HEP), tourism, fishing and food supply, mineral development, and forestry (Figure 14). In addition, rainforests play a vital role in regulating the world's climate and they account for 50 per cent of the world's plants and animals. They are vital too for the protection of soil and water resources.

Industrial uses	Ecological uses	Subsistence uses
Charcoal	Watershed protection	Fuelwood and charcoal
Saw logs	Flood and landslide	Fodder for agriculture
Gums, resins and oils protection		Building poles
Pulpwood	Soil erosion control	Pit sawing and saw milling
Plywood and veneer	Climate regulation	Weaving materials and dyes
Industrial chemicals	e.g. CO_2 and O_2 levels	Rearing silkworms and beekeeping
Medicines	Special woods and ashes	
Genes for crops	Fruit and nuts	
Tourism		

Figure 14 The value of tropical rainforests

The year-round growing season is very attractive for farmers, although the poor quality of the soil results in the land being farmed only for a few years before the land is abandoned (Figure 15). Nevertheless, large-scale plantations occur in areas of tropical rainforests, producing crops such as palm oil, and increasingly being used for the biofuels industry. High rainfall totals, especially in hilly areas, favour the development of HEP, such as at Batang Ai in Sarawak, Malaysia. Areas of rainforest

Figure 15 Shifting cultivation in a rainforest: rice-growing in Sarawak

have a long history of commercial farming. Tropical hardwoods, such as teak and mahogony, are prized by furniture manufacturers. Minerals such as iron ore at Carajas in Brazil and ilminite on the south-east coast of Madagascar are also developed in some rainforest areas.

Hazards in rainforest areas

There are a number of problems in trying to use tropical rainforests for human activities. The removal of vegetation can have many effects, including:

- disruption to the circulation and storage of nutrients
- surface erosion and compaction of soils
- sandification
- increased flood levels and sediment content of rivers
- climatic change
- loss of biodiversity.

Deforestation disrupts the closed system of **nutrient cycling** within tropical rainforests. Nutrients are released through burning and are quickly washed out of the system by the high intensity rains. Thus soils become very infertile very quickly. **Soil erosion** is also associated with vegetation removal (Figure 16). As a result of soil compaction, there is a decrease in infiltration and an increase in overland runoff and soil erosion. **Sandification** is a process of selective erosion. Raindrop impact washes away the finer particles of clay and humus, leaving behind the coarser and heavier sand. Evidence of sandification dates back to the 1890s in Rondonia, Brazil.

As a result of the intense surface runoff and soil erosion, rivers have a higher **flood peak** and a shorter time lag. However, in the dry season river levels are lower, but the rivers carry an increased bed load, and more silt and clay in suspension.

Other changes relate to **climate**. As deforestation progresses, there is a reduction of water that is re-evaporated from the vegetation – hence the recycling of water must diminish. Thus mean annual rainfall is reduced, and the seasonality of rainfall increases.

Soil degradation	3000
Water contamination	1000
Deforestation	750
Coastal erosion	150
Gully erosion	100
Fishery losses	50
Water hyacinth	50
Wildlife losses	10
Total	5110

Figure 17 Annual costs of inaction (US$ million/year)

Figure 16 Rainforest at Batang Ai affected by flooding, shifting cultivation and soil erosion

Case Study

The cost of environmental inaction in Nigeria

As an example, the long-term losses to Nigeria of not acting on growing environmental problems are estimated to be around US$5000 million annually (Figure 17).

Activities

1 Briefly explain two opportunities for developing tropical rainforest areas.

2 Suggest reasons why tourism in tropical rainforests is limited.

3 Explain why the protection of tropical rainforests has global importance.

4 Choose an appropriate method to show the data in Figure 17. Comment on the results you have produced.

5 What are the main hazards associated with the development of tropical rainforests?

▪ Opportunities and hazards in deserts

Land use in desert areas is limited. These areas offer limited potential for agriculture, although with irrigation water, from rivers such as the Nile and the deep aquifers below Libya and south-west USA, farming is possible and profitable (Figure 18). With water, there are plants and the organic content of the soils can develop. In semi-arid areas low-intensity agriculture, such as cattle or sheep ranching, is economically viable without irrigation.

For many areas, such as Tunisia, Jordan and Dubai, tourism is seen as having great potential for economic development, on account of the predictable hot, dry conditions there (Figure 19). At Tabarka, on the northern coast of Tunisia, a new integrated tourist

Figure 18 Irrigation farming in Keiskammahoek, South Africa

Figure 19 A desert area near Teide, Tenerife

route has been created which links the coast with the desert and mountain oasis at Tamerza. This follows the old Arab trading routes and uses accommodation in modern versions of the traditional caravaneserai (hotels or staging posts). For the same reason, arid and semi-arid climates are also proving to be popular areas for retirement homes (the Sun Belt of the south-west of USA, for example). The hot, dry climate of south-west USA has attracted the aviation industry, with the bonus of having clear skies for flying.

On the other hand, there are many hazards associated with desert environments. These include:

- weathering
- flooding of valleys, alluvial fans and plains (playas)
- increased soil erosion and gullying
- surface subsidence due to water abstraction
- deposition of river sediments
- landslides and rockfalls
- dust storms.

Salt weathering is a major hazard in dry areas due to its ability to weaken engineering structures very rapidly. Fluvial hazards are also important because of the use of valleys and floodplains for settlement and farming. Stream flow in dry areas is seasonal, and in some cases erratic. This makes the potential for flooding even greater due to a combination of:

- high velocities
- variable sediment concentrations
- rapid changes in the location of channels.

According to some geographers, erosion is most effective in dry areas. This is because of the relative lack of vegetation. When it rains a high proportion of rain will hit bare ground, compact it, and lead to high rates of overland runoff. By contrast, in much wetter areas, such as rainforests, the vegetation intercepts much of the rainfall and reduces the impact of rainsplash. At the other extreme, areas that are completely dry do not receive enough rain to produce much runoff. Hence it is the areas which have variable rainfall (and a variable vegetation cover) that experience highest rates of erosion and runoff. Moreover, as the type of agriculture changes, the rate of erosion and overland runoff change. Under intense conditions this causes gully formation.

Figure 20 Desertification in South Africa caused by overgrazing by sheep

Figure 21 Desertification in western China

Some desert hazards can have an impact far away. Dust blowing off Africa contributes most of some 2 billion tonnes' worth of dust in the atmosphere each year. The rest originates in Asia, South America, the USA and Australia. This dust has been linked with a series of environmental impacts in the Caribbean including:

Activities

1 Suggest reasons why hot deserts offer limited opportunities for human activities.

2 Explain why flooding can be a hazard in desert locations.

3 Briefly explain one other hazard in a hot desert environment.

- a decline in coral reefs
- the decrease in sea urchins
- outbreaks of disease which struck the region's marine life.

Scientists believe that the dust may contain bacteria, viruses, fungi and chemicals and these can be transported very large distances. They have identified at least 170 different bacteria and 76 types of fungus in airborne dust collected on the Virgin Islands. In St Augustine, Trinidad it has been noticed that more children are admitted to hospital with asthma immediately after a dust cloud had passed. An increase in dust is linked with a greater incidence of asthma.

UNIT
1

Agricultural systems

Agricultural systems

Individual farms and general types of farming can be seen to operate as a **system**. A farm requires a range of **inputs** such as labour and energy so that the **processes** that take place on the farm can be carried out. The aim is to produce the best possible **outputs** such as milk, eggs, meat and crops. A profit will only be made if the income from selling the outputs is greater than expenditure on the inputs and processes. Figure 1 is an input–process–output diagram for a wheat farm.

Different types of agricultural systems (Figure 2, page 108) can be found within individual countries and around the world. The most basic distinctions are between:

- arable, pastoral and mixed farming
- subsistence and commercial farming
- extensive and intensive farming
- organic and non-organic farming.

Arable, pastoral and mixed farming

Arable farms cultivate crops and are not involved with livestock. An arable farm may concentrate on one crop such as wheat or may grow a range of different crops. The crops grown on an arable farm may change over time. For example, if the market price of potatoes increases, more farmers will be attracted to grow this crop. **Pastoral farming** involves keeping livestock such as dairy cattle, beef cattle, sheep and pigs. **Mixed farming** involves cultivating crops and keeping livestock together on a farm.

Subsistence and commercial farming

Subsistence farming is the most basic form of agriculture where the produce is consumed entirely or mainly by the family who work the land or tend the livestock. If a small surplus is produced it may be sold or traded. Examples of subsistence farming are shifting cultivation and nomadic pastoralism

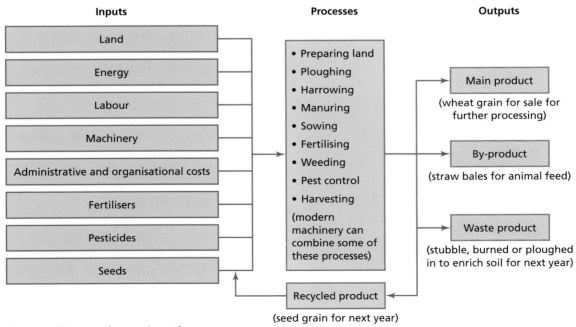

Figure 1 Systems diagram for a wheat farm

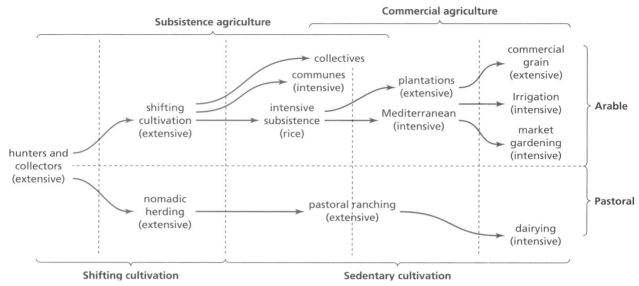

Figure 2 Farming types and levels of development

Figure 3 Arable farming in the Nile valley

Figure 4 Output of local poultry production in a Moroccan market

(Figure 2). Subsistence farming is generally small scale and labour intensive with little or no technological input.

In contrast, the objective of **commercial farming** is to sell everything that the farm produces. The aim is to maximise yields in order to achieve the highest profits possible. Commercial farming can vary from small scale to very large scale.

Extensive and intensive farming

Extensive farming is where a relatively small amount of agricultural produce is obtained per hectare of land, so such farms tend to cover large areas of land. Inputs per unit of land are low. Extensive farming can be both arable and pastoral in nature. In contrast, **intensive farming** is characterised by high inputs per unit of land to achieve high yields per hectare. Examples of intensive farming include market gardening, dairy farming and horticulture. Intensive farms tend to be relatively small in terms of land area.

Organic farming

Organic farming does not use manufactured chemicals and so is practised without the use of chemical fertilisers, pesticides, insecticides and herbicides. Instead animal and green manures are used along with mineral fertilisers such as fish and bone meal. Organic farming therefore requires a higher input of labour than mainstream farming.

Figure 5 Intensive farming – viticulture in the Rioja region of northern Spain

Weeding is a major task with this type of farming. Organic farming is less likely to result in soil erosion and is less harmful to the environment in general. For example, there will be no nitrate runoff into streams and much less harm to wildlife.

Organic farming tends not to produce the 'perfect' potato, tomato or carrot. However, because of the increasing popularity of organic produce it commands a substantially higher price than mainstream farm produce.

Activities

1 Describe the inputs, processes and outputs for the wheat farm shown in Figure 1.

2 **a)** Explain the difference between arable and pastoral farming.
   b) What is mixed farming?

3 Examine the differences between **a)** commercial and subsistence farming and **b)** intensive and extensive farming.

4 Describe the characteristics of organic farming.

Case Study

Plantations: rubber in Malaysia

Plantation crops

Plantations are large farms producing a single cash crop. They were originally developed in tropical areas by European and North American merchants. Large areas were cleared, and a single crop of either a bush or tree was planted. Plantations are an example of monoculture (growing one dominant crop). Plantation crops include tea, coffee, rubber, bananas and sugar cane. Plantations require a high capital and labour input. Some large plantations provide housing, schools and other services for their workers. Local processing facilities may also be present.

History and location

Rubber is one of the main plantation crops in Malaysia. A plantation is defined in Malaysia as an estate over 40 ha. Most rubber is now synthesised from petroleum. However, about a quarter of the world's rubber comes from a natural source, the tropical hevea tree. Rubber was introduced to Malaysia in the late nineteenth century by British entrepreneurs who had smuggled rubber seedlings out of Brazil. The development of the motor car and in particular the need for tyres dramatically increased the demand for rubber and plantations were developed rapidly to satisfy this demand. Large areas of rainforest were cut down to make way for the rubber plantations. Chinese and Indian labour was imported to increase the agricultural labour force. These migrations have been largely responsible for the current ethnic mix of the country.

The hevea tree grows best at average temperatures between 21 and 28 °C, where annual precipitation is just under 2000 mm. Its best growing area is therefore about 10° either side of the equator. However, it is also cultivated further north in China, Guatemala and Mexico. Malaysia, Indonesia and Thailand together produce around 90 per cent of the world's natural rubber. In Malaysia the hevea trees grow best on the gentle

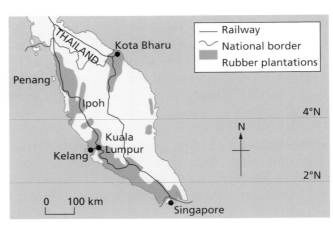

Figure 6 West Malaysia

Figure 7 A rubber tapper at work

lower slopes of the mountains which form the backbone of the Malay peninsula (Figures 6, page 109, and 7). Access to railway lines and the main ports is also a factor in the location of plantations.

Over half of Malaysia's rubber comes from thousands of privately owned smallholdings which average about 2 ha. The rest is cultivated on large estates, owned by the government or various companies. Individual estates can cover over a thousand hectares.

The rubber plantation system

For new planting the seeds are set in raised germination beds about 90 cm wide, with space between the beds to allow room for walking. Tapping begins in the fifth to seventh year after planting, and then continues for 25–30 years. The trees are removed when they are too old and yields begin to decline, and they are replaced with new seedlings. A 30-year-old rubber tree is about 30 m tall. Latex is a milk-like fluid contained in tiny cells beneath the outer bark of the tree. The process of collecting rubber is as follows:

- The tapper cuts away a thin shaving of the bark about 2 mm thick with a special tapping knife. This pierces the cells and the latex oozes slowly out into a collecting cup placed below. Tapping is done early in the day as the latex flows more easily while it is still relatively cool.
- An experienced tapper can cover up to 600 trees in a morning. Many tappers try to find a second job in the afternoon to increase their income.
- Once the tree stem is cut the latex will flow for 1–3 hours, yielding up to three-quarters of a cup of latex.

- The latex is then coagulated in metal pans using a dilute acid.
- The resulting cake of raw rubber is washed to remove any acid and then rolled to remove excess moisture.
- The rubber is then dried on a rack and smoked over a wood fire to stabilise it, before being sold to rubber manufacturers.

In recent years new varieties of rubber trees, created by cross-breeding, have replaced many of the older trees. The new varieties yield as much as ten times more rubber than the traditional trees. This has increased production dramatically on some plantations.

Activities

1 What is a plantation?

2 Look at Figure 6 (page 109). Describe the location of rubber plantations in Malaysia.

3 Describe the physical conditions favourable to the hevea tree.

4 Explain the inputs, processes and outputs of a rubber plantation.

Case Study

Extensive commercial cereals: the Canadian Prairies

Extensive farming

Three-quarters of Canada's farmland lies in the Prairie Provinces of Alberta, Saskatchewan and Manitoba. It is one of the world's largest areas of extensive commercial cereal farming. Here, farms averaging 300 ha stretch for almost 1500 km from east to west. Two or three workers can run a farm by using a range of modern machinery.

The development of wheat cultivation

The hot, dry summers of the Prairies favour wheat production. However feed grains (oats, rye, barley and corn) are also grown. There are also dairy and beef farms scattered throughout the region.

Figure 8 Grain harvesting on the Prairies

Figure 9 Canada's Central Experimental Farm

Wheat was first grown here in the early 1600s. Since then there have been many changes:

- The family farm has grown from a few hand-worked hectares to a large, highly mechanised unit.
- New strains of wheat can withstand lower temperatures and a shorter growing season. This is 20 days less than in the early nineteenth century.
- Today's wheat is resistant to many diseases which once destroyed huge areas of crops.
- Inputs of fertiliser, herbicides and pesticides have steadily increased.
- Ploughing is now little used as it increases soil erosion by making the topsoil more likely to drift. Instead, cultivators cut weeds below the surface without turning the soil.
- A modern combine harvester allows one person to harvest more than 2000 bushels a day. Farmers now spend 10–15 per cent of their income on machinery and equipment, including repairs.

The Canadian Department of Agriculture spends a substantial amount of money on research and development (Figure 9) for wheat and Canada's other main agricultural products to ensure the industry remains productive and competitive in world markets.

Transport to markets

Most of the grain produced in the Prairies is sold in other parts of Canada or abroad. The transport of grains has a set routine:

- Grain is taken by road to a local elevator. This is a wooden structure that can hold around 70,000 bushels. There are 5000 elevators sited at 2000 railway shipping points in the Prairies.
- The amount and quality of grain required at the shipping terminals is decided six weeks before delivery. The terminals are at Thunder Bay, Vancouver, Prince Rupert and Churchill (Figure 10).

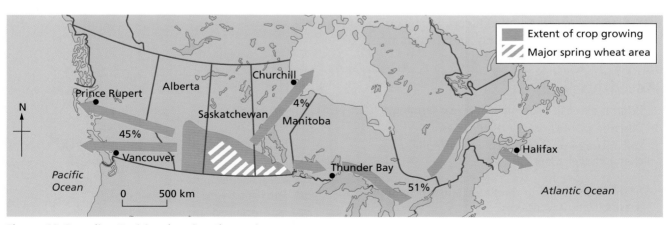

Figure 10 Canadian Prairies showing the grain export routes

- Elevator managers report weekly on the stocks they are holding.
- Shipping instructions are issued on the basis of the 48 loading zones in the Prairies.
- Trains pick up the grain and deliver it to a port. Here it is either loaded on board ship or placed in huge terminal storage elevators.

Physical hazards

At times, severe droughts occur in the Prairies when farmers lose part or all of their crops. While some of the drier areas of the Prairies have good irrigation systems, other areas are not so fortunate. Dry soil crumbles into fine particles which are easily picked up by the wind. In fact, the main long-term problem is soil erosion. Prairie farmers try to protect the soil by:

- Crop–fallow rotation – fields are left fallow every other year to replenish soil moisture. The crop stubble is left to further reduce the effect of the wind.
- Ripping – a caterpillar tractor is used in winter to cut the frozen soil and knurl it into one-third of a metre chunks. These chunks break the effect of the wind close to the soil surface.
- Strip farming – the farmer leaves narrow strips of fallow land, perpendicular to the prevailing wind, between seeded fields.

Export competition and global stocks

Export sales of wheat are important to Prairie farmers. Without exports farm incomes would fall. There is often intense competition between sellers on the world food market. The USA is Canada's biggest competitor in exporting wheat. This has caused arguments between the two countries. Both countries give subsidies to grain farmers which has led to claims of unfair competition. However, the fall in global grain stocks in recent years has put wheat and other grain producers in a very good position as prices have risen and buyers want to purchase all that is produced.

Activities

1 What is extensive commercial cereal farming?

2 Describe the location of the Prairies.

3 Discuss three ways in which wheat farming has changed over time.

4 Look at Figure 10 (page 111). Describe the locations of the grain shipping terminals. Suggest reasons for these different locations.

Case Study

Extensive livestock production: sheep farming in Australia

Characteristics and location

Sheep farming in Australia occupies an area of about 85 million hectares, making it one of Australia's major land uses. Australia is the world's leading sheep producing country with a total of about 120 million sheep. As well as being the largest wool producer and exporter, Australia is also the largest exporter of live sheep and a major exporter of lamb and mutton. The sheep and wool industry is an important sector of Australia's economy.

Sheep are raised throughout southern Australia in moderate to high rainfall areas and in the drier areas of New South Wales and Queensland. Seventy-five per cent of the country's sheep are Merinos which produce very high-quality clothing wool. Merino sheep are able to survive in harsh environments and yet produce heavy fleeces. Sixteen per cent of Australia's sheep are bred for meat production and are a mixture of breeds such as Border Leicester and Dorset. The remaining 9 per cent are a mixture of Merino and cross-bred sheep used for wool and meat production. There are about 60,000 sheep farms in Australia overall, carrying from a few hundred sheep to over 100,000 animals.

Sheep and wool production occurs in three geographical zones (Figure 11):

- high rainfall coastal zone
- wheat/sheep intermediate zone
- pastoral interior zone.

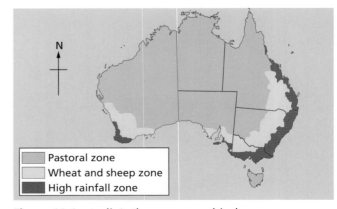

Figure 11 Australia's three geographical zones

About a quarter of all sheep are farmed in the pastoral zone. Sheep farming in Australia in general is extensive in nature but this type of agriculture is at its most extensive in the pastoral zone, which is the arid and semi-arid inland area. Here, summer temperatures are high, rainfall is low and the area is prone to drought. Because of the lack of grass in this inhospitable environment, sheep are often left to eat saltbush and bluebush. In the pastoral zone the density of sheep per hectare is extremely low due to the poor quality of forage. The overall farming input in terms of labour, capital, energy and other inputs is also very low. It is in fact the lowest input per hectare farming in the country. Not surprisingly, farms can be extremely large in terms of area.

In the coastal and intermediate zones the best land is reserved for arable farming, dairy and beef cattle and market gardening. Sheep are frequently kept on the more marginal areas, for example on higher and colder land in the New South Wales highlands, where more profitable types of farming are not viable.

About two-thirds of Australia's sheep are on farms with over 2000 animals. The smallest sheep farms are generally those on the better-quality land, where it is possible to keep many more animals per hectare than in the pastoral zone.

Farming issues

The main issues in Australian sheep farming areas are:
- weed infestation which is difficult to control on very large extensive farms which yield relatively small profits per hectare
- destruction of wildlife habitats due to sheep grazing, particularly in marginal areas

Figure 12 Sheep farm in Australia's pastoral zone

- the occurrence of periodic droughts which make farming conditions even more difficult in low rainfall areas
- soil loss from wind erosion and loss of soil structure – in some areas this is transforming traditional 'mainstream' farming areas into marginal lands
- animal welfare, particularly in the most inhospitable environments where the low human input means that individual animals may not be seen for long periods of time
- increasing concern about the shortage of experienced sheep shearers. Many have left the industry because of poor working conditions and the attraction of better-paid jobs in the mining industry and elsewhere. The number of experienced shearers fell by about a quarter between 2003 and 2006. A good shearer can shear up to 200 sheep in one day.

Activities

1 Why is sheep farming in Australia considered to be 'extensive farming'?

2 Describe the three geographical zones where sheep are kept.

3 Briefly discuss the main issues affecting sheep farming in Australia today.

Case Study

Intensive rice production: the lower Ganges valley

Location

An important area of intensive subsistence rice cultivation is the lower Ganges valley (Figure 13, page 114) in India and Bangladesh. The Ganges Basin is India's most extensive and productive agricultural area and its most densely populated. The delta region of the Ganges occupies a large part of Bangladesh, one of the most densely populated countries in the world. Rice contributes over 75 per cent of the diet in many parts of the region. The physical conditions in the lower Ganges valley and delta are very suitable for rice cultivation:

- temperatures of 21 °C and over throughout the year, allowing two crops to be grown annually (rice needs a growing season of only 100 days)
- monsoon rainfall over 2000 mm, providing sufficient water for the fields to flood, which is necessary for wet rice cultivation
- rich alluvial soils built up through regular flooding over a long time period during the monsoon season
- an important dry period for harvesting the rice.

A water-intensive staple crop

Rice is the staple or main food crop in many parts of Asia. This is not surprising considering its high nutritional value. Current rice production systems are extremely water intensive. Ninety per cent of agricultural water in Asia is used for rice production. The International Rice Research Institute estimates that it takes 5000 litres of water to produce 1 kg of rice. Much of Asia's rice production can be classed as intensive subsistence cultivation where the crop is grown on very small plots of land using a very high input of labour. Rice cultivation by small farmers is sometimes referred to as 'pre-modern intensive farming' because of the traditional techniques used, in contrast to intensive farming systems in MEDCs such as market gardening which are very capital intensive.

'Wet' rice is grown in the fertile silt and flooded areas of the lowlands while 'dry' rice is cultivated on terraces on the hillsides. A **terrace** is a levelled section of a hilly cultivated area. Terracing is a method of soil conservation. It also prevents the rapid runoff of irrigated water. Dry rice is easier to grow but provides lower yields than wet rice.

The farming system

Padi-fields (flooded parcels of land) characterise lowland rice production. Water for irrigation is provided either when the Ganges floods or by means of gravity canals. At first, rice is grown in nurseries. It is then transplanted when the **monsoon rains** flood the padi-fields. The flooded padi-fields may be stocked with fish for an additional source of food. The main rice crop is harvested when the drier season begins in late October. The rice crop gives high yields per hectare. A second rice crop can then be planted in November but water supply can be a problem in some areas for the second crop.

Water buffalo are used for work. This is the only draft animal adapted for life in wetlands. The water buffalo provide an important source of manure in the fields. However, the manure is also used as domestic fuel. The labour-intensive nature of rice cultivation provides work for large numbers of people. This is important in areas of very dense population where there are limited alternative employment opportunities. The low incomes and

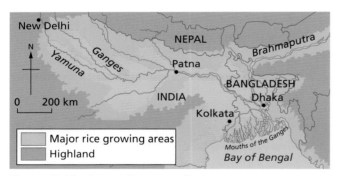

Figure 13 The lower Ganges valley

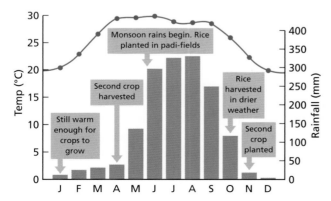

Figure 14 Climate graph for Kolkata

Figure 15 Rice padi-field scene in lower Ganges valley

lack of capital of these subsistence farmers mean that hand labour still dominates in the region. It takes an average of 2000 hours a year to farm 1 ha of land. A high labour input is needed to:

- build the embankments (bunds) that surround the fields; these are stabilised by tree crops such as coconut and banana
- construct irrigation canals where they are required for adequate water supply to the fields
- plant nursery rice, plough the padi-field, transplant the rice from the nursery to the padi-field, weed and harvest the mature rice crop
- cultivate other crops in the dry season and possibly tend a few chickens or other livestock.

Rice seeds are stored from one year to provide the next year's crop. During the dry season when there may be insufficient water for rice cultivation, other crops such as cereals and vegetables are grown. Farms are generally small, often no more than one hectare in size. Many farmers are tenants and pay for use of the land by giving a share of their crop to the landlord.

Activities

1 Describe the location of the lower Ganges valley.

2 Why is rice cultivation in the area considered to be an intensive form of agriculture?

3 Explain why the physical environment provides good conditions for rice cultivation.

4 Describe the inputs, processes and outputs of this type of agriculture.

Case Study

Shifting cultivation in the Amazon Basin

A traditional rainforest system

Shifting cultivation is a traditional farming system that developed a long time ago in tropical rainforests. An area of forest is cleared to create a small plot of land which is cultivated until the soil becomes exhausted. The plot is then abandoned and a new area cleared. Frequently the cultivators work

in a circular pattern, returning to previously used land once the natural fertility of the soil has been renewed. Shifting cultivation is also known as 'slash and burn' and by various local names such as 'chitimene' in Central Africa.

Shifting cultivation can be regarded as a very energy-efficient and sustainable form of forest management. However, as with other traditional forms of farming, very few people are involved in it today compared with a century ago. The exploitation of the rainforests for a variety of commercial reasons, including logging, mining and clearing for ranches, has forced cultivators deeper into the untouched forest areas or onto designated reservations.

In the Amazon rainforest shifting cultivation has been practised for thousands of years by groups of Amerindians who initially had no contact with the outside world. It is likely that there are some groups where this 'isolated' situation still exists, but for most Amerindians there are varying degrees of contact with mainstream Brazilian society.

A fragile ecosystem

Shifting cultivation is necessary in the Amazon rainforest because of the limited fertility of the fragile soils. About a hectare of forest is cleared, although sometimes a few larger trees are left standing to protect crops from the intense heat and heavy rain of the region. Trees providing food such as kola nut and bananas are also left in place. Once the felled vegetation is dry it is burnt. This helps to clear the land, remove weeds and provide ash for

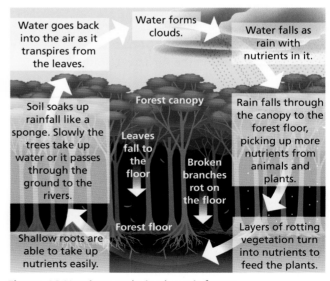

Figure 16 Nutrient cycle in the rainforest

fertiliser. A drawback is that burning also destroys useful organic material and bacteria. Crops such as yams, manioc, beans, tobacco, coca and pumpkins are then planted. The Amerindians also hunt, fish and collect fruit to add to their diet.

When the forest is cleared the nutrient cycle of this ecosystem is broken. Without protection from the dense vegetation, the heavy rain erodes and leaches the soil. This leads to a rapid decline in soil fertility within four or five years. Re-infestation of weeds adds to the problems faced by the shifting cultivators. The only option is to abandon the land and clear another suitable area. The abandoned land gradually regains its fertility, and after about 25 years it may be ready to be farmed again.

Shifting cultivation in the Amazon rainforest is not confined to Amerindians. More recent settlers, many from drought-prone north-east Brazil, have adopted shifting cultivation after migrating to the Amazon region. Often this has not been out of choice. When they came to the Amazon and cleared patches of land, they hoped to farm the land permanently, being unaware that the soil quickly

Figure 19 The River Amazon near Manaus

lost fertility once the nutrient cycle was broken. Figure 17 shows a pair of modern shifting cultivators in their area of land near Itaquatherao, east of Manaus. Figure 18 shows the shack they built. It lacks connection to any modern amenities. A government office about 20 km away tries to look after the welfare of the small farmers who have moved into this area. This includes educating them about the nutrient cycle of the ecosystem. The government office is only linked by a dirt track to the area of shifting cultivators.

A system threatened by deforestation

Shifting cultivation, both traditional and the more recent, has been threatened by the huge land use changes in the Amazon region during the last 40 years or so. Large areas of forest have been lost due to clearance for ranching, mining, logging, road building and the construction of settlements. The number of traditional shifting cultivators has declined considerably.

Figure 17 Cultivators in the Amazon rainforest

Figure 18 House of shifting cultivators

Activities

1 Explain the nutrient cycle in the rainforest.

2 Describe how shifting cultivation operates in the Amazon rainforest.

3 Why can shifting cultivation be regarded as a sustainable form of agriculture?

4 Why are so few people engaged in shifting cultivation today compared with 50 years ago?

Causes and effects of food shortages

Causes

About 800 million people in the world suffer from hunger. The problem is mainly concentrated in Africa but it also has an impact on a number of Asian and Latin American countries. In early 2006 the UN's Food and Agriculture Organisation warned that 27 sub-Saharan countries could need food assistance. Food shortages can occur because of both natural and human problems. The natural problems that can lead to food shortages include:

- soil exhaustion
- drought
- floods
- tropical cyclones
- pests
- disease.

However, economic and political factors can also contribute to food shortages. Such factors include:

- low capital investment
- rapidly rising population
- poor distribution/transport difficulties
- conflict situations.

The impact of such problems has been felt most intensely in LEDCs, where adequate food stocks to cover emergencies affecting food supply usually do not exist. However, MEDCs have not been without their problems. In 2007, the USA faced its worst summer drought since the Dust Bowl years in the 1930s. Other MEDCs such as Australia have also experienced major drought problems. Thus MEDCs are not immune from the physical problems that can cause food shortages. However, they invariably have the human resources to cope with such problems so actual food shortages do not generally occur.

Short-term and long-term effects

The effects of food shortages are both short and longer term. Malnutrition can affect a considerable number of people, particularly children, within a relatively short period when food supplies are significantly reduced. With malnutrition people are less resistant to disease and more likely to fall ill. Such diseases include beri-beri (vitamin B1 deficiency), rickets (vitamin D deficiency) and kwashiorkor (protein deficiency). People who are continually starved of nutrients never fulfil their physical or intellectual potential. Malnutrition reduces people's capacity to work so that land may not be properly tended and other forms of income successfully pursued. This is threatening to lock parts of the developing world into an endless cycle of ill-health, low productivity and underdevelopment.

Case Study

Sudan

In recent years there have been severe food shortages in the Sudan (Figure 20), Africa's largest country. The long civil war and drought have been the main reasons for famine in the Sudan, but there are many associated factors as well (Figure 21, page 118). The civil war, which has lasted for over 20 years, is between the government in Khartoum and rebel forces in the western region of Darfur and in the South. A Christian Aid document in 2004 described the Sudan as 'a country still gripped by a

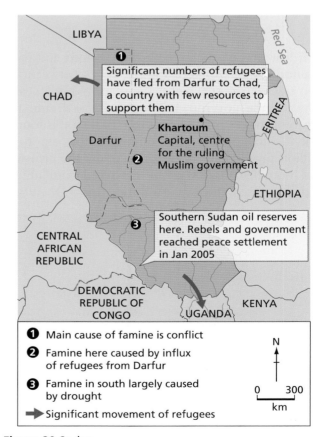

Figure 20 Sudan

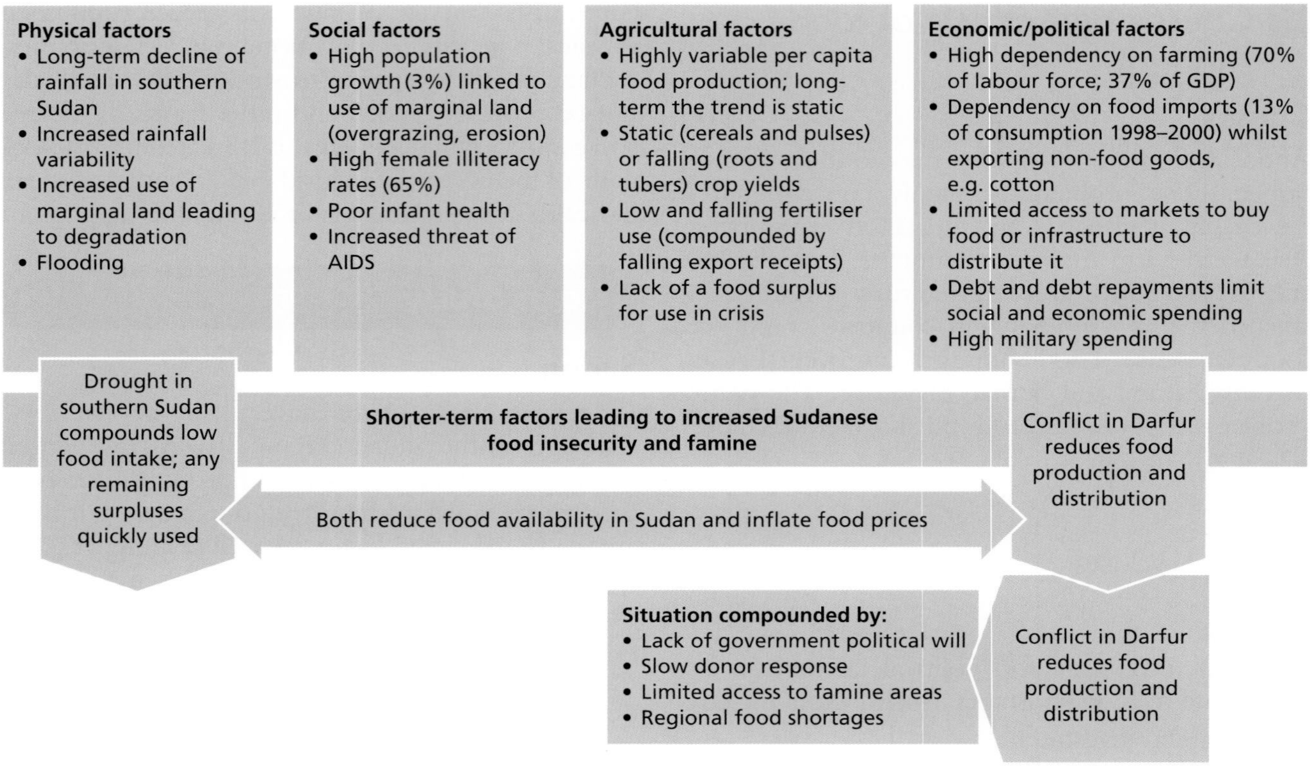

Physical factors	Social factors	Agricultural factors	Economic/political factors
• Long-term decline of rainfall in southern Sudan • Increased rainfall variability • Increased use of marginal land leading to degradation • Flooding	• High population growth (3%) linked to use of marginal land (overgrazing, erosion) • High female illiteracy rates (65%) • Poor infant health • Increased threat of AIDS	• Highly variable per capita food production; long-term the trend is static • Static (cereals and pulses) or falling (roots and tubers) crop yields • Low and falling fertiliser use (compounded by falling export receipts) • Lack of a food surplus for use in crisis	• High dependency on farming (70% of labour force; 37% of GDP) • Dependency on food imports (13% of consumption 1998–2000) whilst exporting non-food goods, e.g. cotton • Limited access to markets to buy food or infrastructure to distribute it • Debt and debt repayments limit social and economic spending • High military spending

Drought in southern Sudan compounds low food intake; any remaining surpluses quickly used

Shorter-term factors leading to increased Sudanese food insecurity and famine

Conflict in Darfur reduces food production and distribution

Both reduce food availability in Sudan and inflate food prices

Situation compounded by:
• Lack of government political will
• Slow donor response
• Limited access to famine areas
• Regional food shortages

Conflict in Darfur reduces food production and distribution

Figure 21 Causes of famine in Sudan

civil war that has been fuelled, prolonged and part-financed by oil'. One of the big issues between the two sides in the civil war is the sharing of oil wealth between the government-controlled north and the south of the country where much of the oil is found. The United Nations has estimated that up to 2 million people have been displaced by the civil war and more than 70,000 people have died from hunger and associated diseases. At times, the UN World Food Programme has stopped deliveries of vital food supplies because the situation has been considered too dangerous for the drivers and aid workers.

Case Study

Zimbabwe

Zimbabwe is another country where there have been food shortages. Again, it has been due to a combination of human and physical problems. However, most experts say the main problem has been poor government decisions. In the early and middle twentieth century the country was a significant exporter of food. However, after Independence the government introduced a programme of land reform. Many of the farms run by people of European descent were seized by the government. The land was subdivided and given to landless people. While some of the motive behind this was greater social and economic equality, the loss of farm management expertise caused food

Figure 22 The fertile banks of the river Nile in Sudan with desert beyond

production to fall sharply. The government argues that the cause lies in the long-running drought. Figure 23 shows the cereal availability situation in Zimbabwe in 2004. There were big contrasts between different areas of the country.

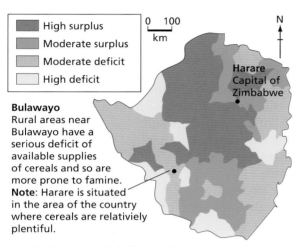

Bulawayo
Rural areas near Bulawayo have a serious deficit of available supplies of cereals and so are more prone to famine.
Note: Harare is situated in the area of the country where cereals are relativiely plentiful.

Figure 23 Cereal availability in Zimbabwe

Activities

1 **a)** With the help of Figures 20 (page 117) and 21 (page 118), explain the causes of food shortages in Sudan.
  b) Suggest what needs to happen for the situation to improve.

2 **a)** Describe the variations in cereal availability in Zimbabwe shown in Figure 23.
  b) Examine the different opinions on the cause of food shortages in Zimbabwe.

◼ Food aid and the green revolution

Food aid

According to the charity ActionAid, there are three types of food aid:

- **Relief food aid** which is delivered directly to people in times of crisis.
- **Programme food aid** which is provided directly to the government of a country for sale on local markets. This usually comes with conditions from the donor country.
- **Project food aid** which is targeted at specific groups of people as part of longer-term development work.

The USA and the EU together provide about two-thirds of global food aid deliveries. At the international level the main organisations are the UN World Food Programme (WFP), the UN Food and Agriculture Organisation (FAO) and the Food Aid Convention.

Food aid is vital to communities in many countries, particularly in Africa but also in parts of Asia and Latin America. However, it is not without controversy:

- The charity CARE has criticised the method of US food aid to Africa. CARE sees the selling of heavily subsidised US produced food in African countries as undermining the ability of African farmers to produce for local markets, making countries even more dependent on aid to avoid famine. CARE wants the USA to send money to buy food locally instead.
- Friends of the Earth say that a genetically modified rice, not allowed for human consumption, originating in the USA has been found in food aid in West Africa.
- Food aid is very expensive, not least because of the high transport costs involved.

There have been recent concerns that food aid may be required for even more people in the future. In 2008, the term 'global food crisis' was being used more and more by the media. Steep increases in the price of food were causing big problems in a number of countries. Major protests about the price of food took place in countries including Haiti, Indonesia, the Philippines and Egypt. In early 2008 the World Bank noted that food prices had risen by 83 per cent in the previous three years. The World Bank warned that progress on development could be destroyed by rapidly rising food costs.

Figure 24 Food aid being delivered in Somalia

The Green Revolution

The package of agricultural improvements generally known as the Green Revolution was seen as the answer to the food problem in many parts of the developing world in the post-1960 period. India was one of the first countries to benefit when a high-yielding variety seed programme (HVP) commenced in 1966–67. In terms of production it was a turning point for Indian agriculture which had virtually reached stagnation. The HVP introduced new hybrid varieties of five cereals: wheat, rice, maize, sorghum and millet. All were drought-resistant with the exception of rice, were very responsive to the application of fertilisers, and had a shorter growing season than the traditional varieties they replaced. Although the benefits of the Green Revolution are clear, serious criticisms have also been made. The two sides of the story can be summarised as follows:

Advantages

- Yields are two to four times greater than traditional varieties.
- The shorter growing season has allowed the introduction of an extra crop in some areas.
- Farming incomes have increased, allowing the purchase of machinery, better seeds, fertilisers and pesticides.
- The diet of rural communities is now more varied.
- Local infrastructure has been upgraded to accommodate a stronger market approach.
- Employment has been created in industries supplying farms with inputs.
- Higher returns have justified a significant increase in irrigation.

Disadvantages

- High inputs of fertiliser and pesticide are required to optimise production. This is costly in both economic and environmental terms. In some areas rural indebtedness has risen sharply.
- HYVs (high-yielding varieties) require more weed control and are often more susceptible to pests and diseases.
- Middle and higher-income farmers have often benefited much more than the majority on low incomes, thus widening the income gap in rural communities. Increased rural to urban migration has often been the result.
- Mechanisation has increased rural unemployment.
- Some HYVs have an inferior taste.
- The problem of salinisation has increased along with the expansion of the irrigated area.

In recent years a much greater concern has arisen about Green Revolution agriculture. In the early 1990s nutritionists noticed that even in countries where average food intake had risen, incapacitating diseases associated with mineral and vitamin deficiencies remained commonplace and in some instances had actually increased. A 1992 UN report directly linked some of these deficiencies to the increased consumption of Green Revolution crops. The problem is that the high-yielding varieties introduced during the Green Revolution are usually low in minerals and vitamins. Because the new crops have displaced the local fruits, vegetables and legumes that traditionally supplied important vitamins and minerals, the diet of many people in the developing world is now extremely low in zinc, iron, vitamin A and other micronutrients.

The Green Revolution has been a major factor enabling global food supply to keep pace with population growth, but with growing concerns about a new food crisis new technological advances may well be required to improve the global food security situation.

Figure 25 Green Revolution crops being harvested in Brazil

Activities

1 Describe the different types of food aid.

2 Why is food aid sometimes controversial?

3 Discuss the advantages and disadvantages of Green Revolution farming.

Industrial systems

Classification of industry

In all modern economies of a significant size, people do hundreds – and in some cases thousands – of different jobs, all of which can be placed into three broad economic sectors:

- **The primary sector** (Figure 1) exploits raw materials from land, water and air. Farming, fishing, forestry, mining and quarrying make up most of the jobs in this sector. Some primary products are sold directly to the consumer but most go to secondary industries for processing.
- **The secondary sector** (Figure 2) manufactures primary materials into finished products. Activities in this sector include the production of processed food, furniture and motor vehicles. Secondary products are classed either as consumer goods (produced for sale to the public) or capital goods (produced for sale to other industries).

- **The tertiary sector** (Figure 3) provides services to businesses and to people. Retail employees, drivers, architects, teachers and nurses are examples of occupations in this sector.

As an economy advances the proportion of people employed in each sector changes (Figure 4, page 122). Countries such as the USA, Japan, Germany and the UK are 'post-industrial societies' where the majority of people are employed in the tertiary sector. Yet in 1900 40 per cent of employment in the USA was in the primary sector. However, the mechanisation of farming, mining, forestry and fishing drastically reduced the demand for labour in these industries. As these jobs disappeared people moved to urban areas where secondary and tertiary employment was expanding. Less than 4 per cent of employment in the USA is now in the primary sector.

Figure 1 Primary industry: an oil well in Dorset, UK

Figure 2 Secondary industry: grain processing factories in Chicago, USA

Figure 3 Tertiary industry: a street market in Nabul, Tunisia

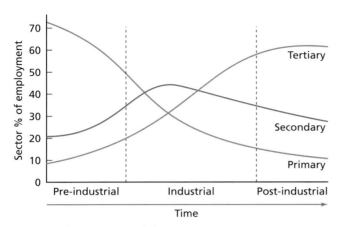

Figure 4 The sector model

Human labour is steadily being replaced in manufacturing too. In more and more factories, robots and other advanced machinery handle assembly-line jobs which once employed large numbers of people. In 1950 the same number of Americans were employed in manufacturing as in services. By 1980 two-thirds were working in services. Today 78 per cent of Americans work in the tertiary sector.

The tertiary sector is also changing. In banking, insurance and many other types of businesses, computer networks have reduced the number of people required. But elsewhere service employment is rising such as in health, education and tourism.

People in the poorest countries of the world are heavily dependent on the primary sector for employment. Such countries are often **primary product dependent** because they rely on one or a small number of primary products for all their export earnings. In **newly industrialised countries** employment in manufacturing has increased rapidly in recent decades. Figure 5 compares the employment structure of an LEDC, an NIC and an MEDC.

Country	% primary	% secondary	% tertiary
Australia (MEDC)	4	21	75
Malaysia (NIC)	13	36	51
Bangladesh (LEDC)	63	11	26

Figure 5 Employment structure of an LEDC, NIC and MEDC

The formal and informal sectors of employment

Jobs in the **formal sector** will be known to the government department responsible for taxation and to other government offices. Such jobs generally provide better pay and much greater security than jobs in the **informal sector**. Fringe benefits such as holiday and sick pay may also be available. Formal sector employment includes health and education service workers, government workers, and people working in established manufacturing and retail companies.

In contrast, the informal sector is that part of the economy operating outside official recognition. Employment is generally low paid and often temporary and/or part-time in nature. While such employment is outside the tax system, job security will be poor with an absence of fringe benefits. About three-quarters of those working in the informal sector are employed in services. Typical jobs are shoe-shiners, street food stalls, messengers, repair shops and market traders. Informal manufacturing tends to include both the workshop sector, making cheap furniture for example, and the traditional craft sector. Many of these goods are sold in bazaars and street markets.

Figure 6 Informal sector employment: Cairo, Egypt

Activities

1 Look at Figure 4. Describe and explain how the employment structure of MEDCs has changed over time.

2 Explain the difference between the formal and informal sectors of employment.

Industrial systems

Manufacturing industry as a whole or an individual factory can be regarded as a system. Industrial systems, like agricultural systems, have inputs, processes and outputs (Figure 7).

- **Inputs** are the elements that are required for the processes to take place (Figure 8). Inputs include raw materials, labour, energy and capital.
- **Processes** are the industrial activities that take place in the factory to make the finished product (Figure 9). For example, in the car industry processes include moulding sheet steel into the shaped panels that make up the car, welding and painting.
- **Outputs** comprise the finished product or products that are sold to customers. Sometimes **by-products** may be produced. A by-product is something that is left over from the main production process which has some value and therefore can be sold. All manufacturing industries produce **waste product** which has no value and must be disposed of. Costs will be incurred in the disposal of waste product.

Industries are also described or classified by the use of opposing terms such as **heavy industry** and **light industry** (Figure 10, page 124). In this case iron and steel would be an example of a heavy industry, using large amounts of bulky raw materials, processing on a huge scale and producing final products of a significant size. In contrast, the assembly of computers is a light industry.

Figure 8 An input – sugar cane from Brazil

Figure 9 Processing – sugar refinery in Canada

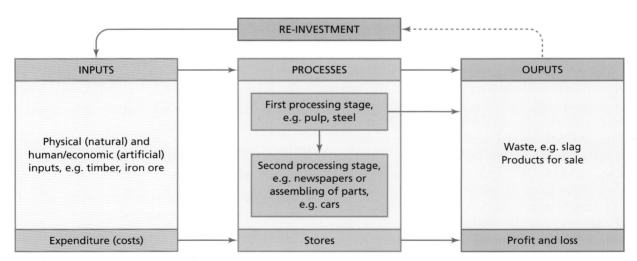

Figure 7 Industrial systems diagram

Classification contrasts	Characteristics
Large scale and small scale	depending on the size of plant and machinery, and the numbers employed
Heavy and light	depending on the nature of processes and products in terms of unit weight
Market oriented and raw material oriented	where the location of the industry or firm is drawn either towards the market or the raw materials required, usually because of transportation costs
Processing and assembly	the former involving the direct processing of raw materials with the latter putting together parts and components
Capital intensive and labour intensive	depending on the ratio of investment on plant and machinery to the number of employees
Fordist and flexible	Fordist industries, named after the assembly-line methods used in the early automobile industry, mass produce on a large scale making standardised products. Flexible industries make a range of specialised products using high technology to respond quickly to changes in demand.
National and transnational	Many firms in the small to medium-size range manufacture in only one country. Transnationals, which are usually extremely large companies, produce in at least two countries but may manufacture in dozens of nations.

Figure 10 Industrial classifications

Activities

1 Explain the industrial systems diagram shown in Figure 7 (page 123).

2 With regard to manufacturing industry explain the difference between:

   a) heavy and light
   b) market oriented and raw material oriented
   c) capital intensive and labour intensive.

◼ Factors affecting the location of industry

Everyday decisions are made about where to locate industrial premises, ranging from small workshops to huge industrial complexes. In general, the larger the company the greater the number of real alternative locations available. For each possible location a wide range of factors can have an impact on total costs and thus influence the decision-making process. The factors affecting industrial location differ from industry to industry and their relative importance is subject to change over time. These factors can be broadly sub-divided into physical and human (Figure 11).

Case Study

◼ High-technology industry: Ottawa and Bangalore

High-technology industry is the fastest-growing manufacturing industry in the world. It all began in Silicon Valley (the Santa Clara valley) south of San Francisco, in the 1960s. Since then it has spread across the world. Virtually all MEDCs and NICs have at least one high-technology cluster (companies grouped together in one region).

High-tech companies use or make silicon chips, computers, software, robots, aerospace components and other technically advanced products. These companies put a great deal of money into scientific research. Their aim is to develop newer and even more advanced products.

'Silicon Valley North': the Ottawa Region of Canada

High-technology is the fastest-growing industry in Canada today. Ottawa has the greatest number of electronics and computer companies in Canada. The area is known as Silicon Valley North (Figure 12), after the world-renowned Silicon Valley

Physical factors	Human factors
Site The availability and cost of land is important. Large factories in particular need flat, well-drained land on solid bedrock. An adjacent water supply may be essential for some industries.	*Capital (money)* Business people, banks and governments are more likely to invest money in some areas than others.
Raw materials Industries requiring heavy and bulky raw materials which are expensive to transport will generally locate as close to these raw materials as possible.	*Labour* Increasingly it is the quality and cost of labour rather than the quantity that are the key factors here. The reputation, turnover and mobility of labour can also be important.
Energy At times in the past, industry needed to be located near fast-flowing rivers or coal mines. Today, electricity can be transmitted to most locations. However, energy-hungry industries, such as metal smelting, may be drawn to countries with relatively cheap hydro-electricity such as Norway.	*Transport and communications* Transport costs are lower in real terms than ever before but remain important for heavy, bulky items. Accessibility to airports, ports, motorways and key railway terminals may be crucial factors for some industries.
Natural routeways and harbours These were essential factors in the past and are still important today as many modern roads and railways still follow natural routeways. Natural harbours provide good locations for ports and the industrial complexes often found at ports.	*Markets* The location and size of markets is a major influence for some industries.
Climate Some industries such as aerospace and film benefit directly from a sunny climate. Indirect benefits such as lower heating bills and a more favourable quality of life may also be apparent.	*Government influence* Government policies and decisions can have a big direct and indirect impact on the location of industry. Governments can encourage industries to locate in certain areas and deny them planning permission in others.
	Quality of life Highly skilled personnel who have a choice about where they work will favour areas where the quality of life is high (leisure facilities, good housing, attractive scenery, etc.).

Figure 11 Factors affecting industrial location

(South) in California. The other main high-technology clusters in Canada are in Toronto, Vancouver and Montreal. Ottawa is a popular location for high-tech industries for several reasons:

- It is the capital city. Decision-makers and experts of all kinds work there.
- The region has benefited from the growth of public scientific research laboratories such as the Atomic Energy of Canada Laboratory.
- Ottawa's two universities are famous for science and engineering. There are strong formal and informal links between high-tech firms and the universities.
- Many high-tech companies began with the help of government grants and many now rely on government contracts for work.
- High-tech firms benefit from the 'agglomeration economies' of being close together. They 'swap' ideas and staff.

Figure 12 Location of Ottawa

- The region has had good access to venture capital which has been vital in providing the start-up costs of many new companies.
- It is close to major Canadian and US markets where many of the products are sold.
- Ottawa is a very pleasant place to live. This helps firms to attract highly skilled workers.
- The city's economic development corporation advertises the advantages of Ottawa internationally.

High-tech companies in the Ottawa region include 3M, Cisco Systems, Dell, Hewlett-Packard, IBM, Intel, LogicVision, Mitel and Nortel. A 2006 survey recorded 1841 high-tech companies in the Ottawa region, employing over 78,000 people (Figure 13). There is a good spread of large, medium-sized and small high-tech companies in the region.

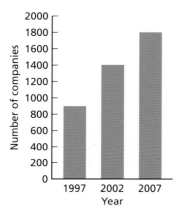

Figure 13 Growth of high-tech industries in the Ottawa region, 1997, 2002 and 2007

Bangalore: India's high-tech city

Bangalore, Hyderabad and Chennai, in the south, along with the western city of Pune and the capital city Delhi, have emerged as the centres of India's high-technology industry.

Bangalore is the most important individual centre in India for high-technology. The city's pleasant climate, moderated by its location on the Deccan Plateau over 900 m above sea level, is a significant attraction to foreign and domestic companies alike. Known as the 'Garden City', Bangalore claims to have the highest quality of life in the country. Because of its dust-free environment, large public-sector undertakings such as Hindustan Aeronautics Ltd and the Indian Space Research Organisation were established in Bangalore by the Indian government. In addition, the state government has a long history of support for science and technology.

The city prides itself on a 'culture of learning' which gives it an innovative leadership within India.

In the 1980s Bangalore became the location for the first large-scale foreign investment in high technology in India when Texas Instruments selected the city above a number of other possibilities. Other multinationals soon followed as the reputation of the city grew. Important backward and forward linkages were steadily established over time. Apart from ICT industries, Bangalore is also India's most important centre for aerospace and biotechnology.

India's ICT sector has benefited from the filtering down of business from the developed world. Many European and North American companies which previously outsourced their ICT requirements to local companies are now using Indian companies. Outsourcing to India occurs because:

- labour costs are considerably lower
- a number of developed countries have significant ICT skills shortages
- India has a large and able English-speaking workforce (there are about 50 million English-speakers in India).

Since 1981, the city's population has more than doubled, from 2.4 million to over 5 million, while the number of vehicles has grown even faster, from fewer than 200,000 cars and scooters to over 1.6 million. The city's landscape has changed dramatically with many new glass-and-steel skyscrapers and numerous cybercafés.

Figure 14 Location of Bangalore

- Location of 925 software companies employing more than 80,000 ICT workers. Bangalore accounts for nearly 40% of India's software exports.

- The city has 46 ICT design companies, 166 systems software companies and 108 communications software companies. Over 40 per cent of Bangalore's software exports are in these high-technology areas.

- Major companies include Tata Consulting Services (TCS), Infosys Technologies, Wipro and Kshema Technologies.

- The 1.7 million m² International Tech Park was set up as a joint venture between the Government of Karnataka, the Government of Singapore and the House of Tata. The Electronic City is an industrial area with over 100 electronics companies such as Infosys, Wipro, Siemens, Motorola, ITI, etc.

- The city has attracted outsourcing right across the ICT spectrum from software development to ICT enabled services.

- The city boasts 21 engineering colleges.

- NASDAQ, the world's biggest stock exchange with a turnover of over $20 trillion, opened its third international office in Bangalore in 2001.

Figure 15 Bangalore factfile

Activities

1 What is high-tech industry?

2 Why do high-tech companies frequently locate close to major universities?

3 What other factors influence the location of such companies?

Case Study

Slovakia: the car industry expands rapidly

Slovakia (Figure 16), and in particular the region around Bratislava, is proving to be one of the main focal points for foreign direct investment in the ten new member countries who joined the EU in May 2004. The car industry, one of the world's major manufacturing industries, is at the forefront of the new investment.

In the Bratislava area skilled labour is abundant because of the former industrial background of the region. However, new investment has not been confined to the capital alone – a number of other locations have also proved attractive to foreign companies. The business magazine *Forbes* has referred to the country as 'an investment paradise', forecasting that this small nation will emulate the previous successes of Hong Kong and the Republic of Ireland. Both are equally small economies, each witnessing very rapid growth in a relatively short period of time. Slovakia has arguably the lowest business cost base of the new EU members.

Location factors

The location factors which have attracted the car industry to Slovakia are:

- relatively low labour costs
- low company taxation rates
- a highly skilled workforce, particularly in areas that were once important for heavy industry
- a strong work ethic resulting in high levels of productivity
- low transport costs because of proximity to western European markets
- very low political risk because of the stable nature of the country
- attractive government incentives due to competition between Slovakia and other potential receiving countries

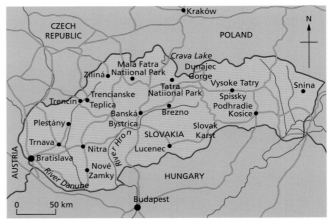

Figure 16 Slovakia

- good infrastructure in and around Bratislava and other selected locations
- an expanding regional market for cars as per capita incomes increase.

Volkswagen expands

Slovakia, with a population just over 5 million, has been dubbed 'the Detroit of Europe' by at least one writer. Prior to EU membership Slovakia already boasted a Volkswagen plant with an output of 250,000 cars a year. The Bratislava plant is one of the top three Volkswagen factories in the world producing the Polo, the Touareg and the SEAT Ibiza. The Touareg is almost totally produced in Slovakia. In 2004 Volkswagen produced almost a quarter of all Slovakia's exports. Although Volkswagen did not receive a subsidy from the Slovak government, the company is benefiting from a ten-year tax concession. A VW spokesperson has been quoted as saying: 'Volkswagen voted for Bratislava because of the good infrastructure of this region – highway, railway, airport and river transport'. Bratislava has a harbour on the River Danube with freight and passenger connections to Vienna and Budapest. Slovakia enjoys a strategic location on the border with Austria.

In addition to its car manufacturing plant in Bratislava, which was founded in 1991, VW also has a plant manufacturing components in Martin. The latter was opened in 2000. Between the two plants, VW employs 8700 workers. In 2001 VW founded the APP Lozorno industrial park. Located 15 km away from the Bratislava plant, it is from here that suppliers such as Brose provide just-in-time deliveries. Other key suppliers which have been attracted to Slovakia by VW include Dura, Kuster and Tower Automotive.

Other recent investment

In late 2004 Slovakia fended off fierce competition from Poland and Hungary to seal a deal with the Korean company Hyundai, to build its first European car factory in the country. The factory opened in 2006 and will produce up to 200,000 vehicles a year under Hyundai's Kia brand. The location of the factory is near Zilina, 200 km north-east of Bratislava. Some of Slovakia's east European competitors have criticised the amount of the Slovak subsidy to Kia – 228 million euros. In addition the state is paying 1750 euros for each of the 3000 workers. The total cost of Kia's investment will be $870 million. As with other large car plants, Kia is attracting some of its main suppliers to locate nearby. With its seven suppliers the total investment is estimated to be $1.4 billion.

In 2006, Peugeot opened a large new car plant in Trnava, 50 km from Bratislava. When it reaches maximum production this state-of-the-art plant will export 300,000 cars a year to western Europe and to other parts of the world. This greenfield investment will cost around 700 million euros. Apart from low wage costs, a major attraction of this location to Peugeot was the promise of free land. The new Peugeot factory is attracting many of its suppliers to the same location. In total, it is

Figure 17 A VW car plant in Bratislava

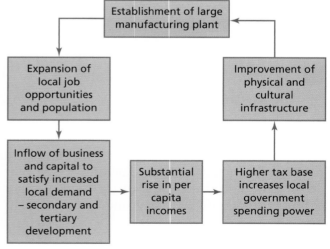

Figure 18 Simplified model of cumulative causation

estimated that 10,000 new jobs will be created. The Slovak Investment and Trade Development Agency (SARIO), established in 2001, played an important role in attracting Peugeot Citroen to the country. SARIO offered a range of potential sites for the new factory.

Cumulative causation

Slovakia produced 571,000 cars in 2007, up from 285,000 the previous year. This makes it the largest per capita car producer in the world. As the pool of skilled labour expands, the likelihood of attracting more foreign motor manufacturers increases. A process of cumulative causation is underway whereby successful economic growth is leading to further expansion of the economy (Figure 18).

Activities

1 Discuss the reasons for the high level of investment in car manufacturing in Slovakia by foreign TNCs.

2 Explain how the process of cumulative causation has been important in improving the standard of living in Slovakia.

Leisure activities and tourism

The growth of leisure and tourism

Over the last 50 years **tourism** has developed into a major global industry which is still expanding rapidly (Figure 1). It is one of the major elements in the process of **globalisation**.

Tourism is defined as travel away from the home environment:

- for leisure, recreation and holidays
- to visit friends and relations
- for business and professional reasons.

Recreation is the use of leisure time for relaxation and enjoyment which does not involve travel away from home.

The development of tourism

The medical profession was largely responsible for the growth of taking holidays away from home. During the seventeenth century doctors increasingly began to recommend the benefits of mineral waters and by the end of the eighteenth century there were hundreds of spas in Britain (Figure 2). Bath and Tunbridge Wells were among the most famous. The second stage in the development of holiday locations was the emergence of the seaside resort. Sea bathing is usually said to have begun in Britain at Scarborough about 1730.

The annual holiday, away from work, for the masses was a product of the Industrial Revolution, which brought big social and economic changes. However, until the latter part of the nineteenth century only the very rich could afford to take a holiday away from home.

The first **package tours** were arranged by Thomas Cook in 1841. These took travellers from Leicester to Loughborough, 19 km away, to attend temperance (abstinence from alcoholic drink) meetings. At the time it was the newly laid railway

Figure 2 The historical mineral waters in the spa town of Bath

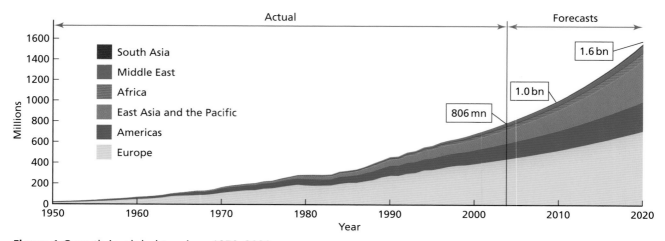

Figure 1 Growth in global tourism, 1950–2020

Economic
■ Steadily rising real incomes
■ Decreasing real costs of holidays
■ Widening range of destinations within the middle-income range
■ Heavy marketing of shorter foreign holidays aimed at those who have the time and disposable income to take an additional break
■ Expansion of budget airlines
■ 'Air miles' and other retail reward schemes aimed at travel and tourism
■ 'Globalisation' has increased business travel considerably

Social
■ Increase in average number of days' paid leave
■ Increasing desire to experience different cultures and landscapes
■ Raised expectations of international travel with increasing media coverage of holidays, travel and nature
■ High levels of international migration over the last decade or so means that more people have relatives and friends living abroad

Political
■ Many governments have invested heavily to encourage tourism
■ Government backing for major international events such as the Olympic Games and the World Cup

Figure 3 Factors affecting global tourism

network that provided the transport infrastructure for Cook to expand his tour operations. Of equal importance was the emergence of a significant middle-class with time and money to spare for extended recreation. It was not long before such activities spread to other countries.

By far the greatest developments have occurred since the end of the Second World War, arising from the substantial growth in leisure time, affluence and mobility enjoyed in developed countries. However, it took the jet plane to herald the era of international mass tourism. In 1970 when Pan Am flew the first Boeing 747 from New York to London, scheduled planes carried 307 million passengers. By 2006 the number had reached 2.1 billion.

Reasons for the growth of global tourism

Figure 3 shows the range of factors responsible for the growth of global tourism. More and more people have become aware of the attractions of the physical and human landscape in their own countries and abroad, and rising living standards have allowed an increasing number of people to experience such attractions.

Recent data

In 2005 **international tourist arrivals** exceeded 800 million, an all-time record (Figure 1). The World Tourism Organisation (WTO) forecasts an increase to 1 billion arrivals in 2010 and 1.6 billion in 2020. **International tourism receipts** totalled $680 billion in 2005 with 70 countries earning more than $1 billion from international tourism.

Fifty per cent of inbound tourism is for the purpose of leisure, recreation and holidays. The second most important reason is visiting friends and relatives. Seasonality is the major problem with tourism as a source of employment, having a major impact on incomes and the quality of life in the less popular times of the year (Figure 4).

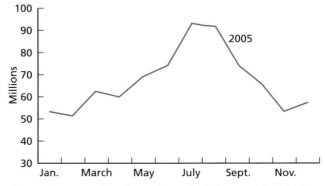

Figure 4 International tourist arrivals by month, 2005

People from MEDCs still dominate global tourism but many emerging economies have shown very fast growth rates in recent years. When people can afford to travel they usually do. **Tourist-generating countries** have a big impact on the flow of money around the world.

The growth of leisure facilities

Leisure facilities have expanded rapidly in both urban and rural areas in the richer countries of the world and in key tourism destinations in poorer countries. Some leisure facilities are publicly owned and run, but an increasing number are operated by private companies. At the upper end of the market, leisure is very big business indeed.

Leisure activities may be classified in the following ways:
- active/passive – depending on the amount of physical exercise involved
- formal/informal – depending on the level to which participation is organised
- resource based/user oriented – depending on the degree of reliance on the natural environment, purpose-built facilities and special equipment.

Theme parks: the largest-scale leisure facilities

Theme parks create artificial destinations from scratch. The largest are located close to Tokyo, Paris, Los Angeles and Orlando. These four Disney theme parks are the world's biggest human facility draws. A vital element of the Disney empire is its research centre in Glendale, California where the very latest technology is used to produce even more exciting novelties and experiences for visitors. It

Figure 5 The Bird's Nest Stadium, Beijing 2008

Figure 6 Inside Manchester United's Old Trafford Stadium – watching football is one of the world's major leisure activities

researches what audiences want and then goes and creates it.

The impact of leisure

The steadily rising demand for leisure affects society in the following ways:
- Spatial – more space is required as demand for facilities in both urban and rural areas increases. Competition for land is intense in many areas.
- Environmental – some activities and the traffic generated by them can do considerable harm to the environment. This is often greatest in rural areas, reaching a maximum at **honeypot** locations.
- Economic – recreation is a significant sector of many national economies. The industry is a big employer but much of the work is seasonal.
- Social – some activities cost little or nothing while others are only open to people on high incomes. Public facilities are often worst in poor areas while those on low incomes can't afford the membership fees of private clubs.

Activities

1 Describe the growth in international tourism shown in Figure 1 (page 130).

2 Discuss the reasons for the growth of tourism.

3 Describe the seasonal nature of tourism as shown in Figure 4 (page 131).

4 Explain the impact of the expansion of leisure facilities in many areas on people and the environment.

■ The benefits and disadvantages of tourism to receiving areas

The economic impact

Many countries, both developed and developing, have put a high level of capital investment into tourism. This is money invested in hotels, attractions, airports, roads and other aspects of infrastructure that facilitates high volume tourism. There has been considerable debate about the wisdom of such a strategy. Do the economic benefits outweigh the economic costs? The majority of people concerned with the tourist industry think they do, but critics of the impact of tourism have presented some strong arguments of their own. Figure 7 shows that tourism has many indirect as well as direct effects.

Supporters of the development potential of tourism argue as follows:

- It is an important factor in the balance of payments of many nations. Tourism brings in valuable foreign currency. This foreign currency is necessary for countries to pay for the goods and services they import from abroad. Many small developing countries have few other resources that they can use to obtain foreign currency.
- Tourism benefits other sectors of the economy, providing jobs and income through the supply chain. It can set off the process of cumulative causation whereby one phase of investment can trigger other subsequent phases of investment.

- It provides governments with considerable tax revenues which help to pay for education, health and other things for which a government has to find money.
- By providing employment in rural areas it can help to reduce rural–urban migration. Such migration is a major problem in many developing countries.
- A major tourism development can act as a **growth pole**, stimulating the economy of the larger region.
- It can create openings for small businesses such as taxi firms, beach facility hire companies and small cafés.
- It can support many jobs in the informal sector, which plays a major role in the economy of many developing countries.

However, critics argue that the value of tourism is often overrated for the following reasons:

- **Economic leakages** (Figure 9, page 134) from developing to developed countries run at a rate of between 60 and 75 per cent. Economic leakages are that part of the money a tourist pays for a foreign holiday that does not benefit the destination country because it goes elsewhere. With cheap package holidays, by far the greater part of the money paid stays in the country where the holiday was purchased.

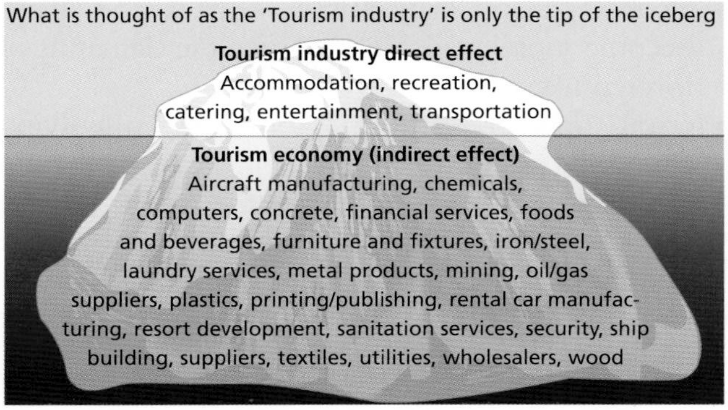

What is thought of as the 'Tourism industry' is only the tip of the iceberg

Tourism industry direct effect
Accommodation, recreation, catering, entertainment, transportation

Tourism economy (indirect effect)
Aircraft manufacturing, chemicals, computers, concrete, financial services, foods and beverages, furniture and fixtures, iron/steel, laundry services, metal products, mining, oil/gas suppliers, plastics, printing/publishing, rental car manufacturing, resort development, sanitation services, security, ship building, suppliers, textiles, utilities, wholesalers, wood

Figure 7 Direct and indirect economic impact of the tourist industry

Figure 8 Beach artist, Agadir, Morocco – an example of informal sector employment

Figure 9 Diagram of economic leakages

Within the diagram:

LEDC TOURIST DESTINATION
Total money spent on tourism to this destination

Transport costs paid to airlines and other carriers

Payments to foreign owners of hotels and other facilities

The cost of goods and services imported for the tourist industry

Remittances sent home by foreign workers

Foreign debt relating to tourism

Payments to foreign companies to build tourist infrastructure

LEAKAGES

Figure 10 Cruise ship on the River Nile – tourism is Egypt's main source of foreign currency

- Tourism is labour intensive, providing a range of jobs, especially for women and young people. However, most local jobs created are menial, low-paid and seasonal. Overseas labour may be brought in to fill middle and senior management positions.
- Money borrowed to invest in the necessary infrastructure for tourism increases the national debt.
- At some destinations tourists spend most of their money in their hotels with minimum benefit to the wider community.
- Tourism might not be the best use for local resources which could in the future create a larger **multiplier effect** if used by a different economic sector.
- Locations can become over-dependent on tourism which causes big problems if visitor numbers fall.
- The tourist industry has a huge appetite for resources which often impinge heavily on the needs of local people. A long-term protest against tourism in Goa highlighted the fact that one five-star hotel consumed as much water as five local villages, and the average hotel resident used 28 times more electricity per day than a local person.
- International trade agreements such as the General Agreement on Trade in Services (GATS) allow the global hotel giants to set up in most countries. Even if governments favour local investors there is little they can do.

The social and cultural impact

The traditional cultures of many communities in the developing world have suffered because of the development of tourism. The disadvantages include the following:

- loss of locally owned land as tourism companies buy up large tracts of land in the most scenic and accessible locations
- abandonment of traditional values and practices
- displacement of people to make way for tourist developments
- changing community structure – communities that were once very close socially and economically may be weakened considerably due to a major outside influence such as tourism
- abuse of human rights by large companies and governments in the quest to maximise profits
- alcoholism and drug abuse as drink and drugs become more available to satisfy the demands of foreign tourists
- crime and prostitution, sometimes involving children ('sex tourism' is a big issue in certain locations such as Bangkok, but it is also present in some degree in most locations visited by large numbers of international tourists)
- visitor congestion at key locations, hindering the movement of local people
- denying local people access to beaches to provide 'exclusivity' for visitors
- loss of housing for local people as more visitors buy second homes in popular tourist areas (Figure 11).

Figure 11 Entrance to a National Park in Andalucia, Spain. The graffiti refers to the number of foreigners buying up houses in the nearby village of Frigiliana.

Figure 12 shows how the attitudes to tourism can change over time. An industry which is usually seen as very beneficial initially can eventually become the source of considerable irritation, particularly where there is a big clash of cultures.

However, tourism can also have positive social and cultural impacts:

- Tourism development can increase the range of social facilities for local people.
- It can lead to greater understanding between people of different cultures.

1	**Euphoria**
	• Enthusiasm for tourist development
	• Mutual feeling of satisfaction
	• Opportunities for local participations
	• Flows of money and interesting contacts
2	**Apathy**
	• Industry expands
	• Tourists taken for granted
	• More interest in profit making
	• Personal contact becomes more formal
3	**Irritation**
	• Industry nearing saturation point
	• Expansion of factilities required
	• Encroachment into local way of life
4	**Antagonism**
	• Irritations become more overt
	• The tourist is seen as the harbinger of all that is bad
	• Mutual politeness gives way to antagonism
5	**Final level**
	• Environment has changed irreversibly
	• The resource base has changed and the type of tourist has also changed
	• If the destination is large enough to cope with mass tourism it will continue to thrive

Figure 12 Doxey's index of irritation caused by tourism

- Family ties may be strengthened by visits to relatives living in other regions and countries.
- Visiting ancient sites can develop a greater appreciation of the historical legacy of host countries.
- It can help develop foreign language skills in host communities.
- It may encourage migration to major tourist-generating countries
- A multitude of cultures congregating together for major international events such as the Olympic Games can have a very positive global impact.

The tourist industry and the various scales of government in host countries have become increasingly aware of the problems the industry creates. They are now using a range of management techniques in an attempt to mitigate such effects. Education is the most important element so that visitors are made aware of the most sensitive aspects of the host culture.

Figure 13 Armed tourism police on a Nile cruise ship

Activities

1 Compare the direct and indirect effects of tourism.

2 Explain how economic leakages occur.

3 Explain the sequence of changes illustrated in Doxey's index (Figure 12).

4 Research the social impact of international tourism in one destination.

Case Study

▇ Jamaica: the benefits and disadvantages associated with the growth of tourism

Economic importance

Tourism has become an increasingly vital part of Jamaica's economy in recent decades. The contribution of tourism to total employment and GDP has risen substantially. It has brought considerable opportunities to the country although it has not been without its problems. Jamaica has been determined to learn from the 'mistakes' of other countries and ensure that the population would gain real benefits from the growth of tourism.

Tourism's direct contribution to GDP in 2007 amounted to almost $1.2 billion. Adding all the indirect economic benefits increased the figure to almost $3.8 billion, or 31.1 per cent of total GDP. Direct employment in the industry amounted to 92,000 but the overall figure which includes indirect employment is over three times as large. In the most popular tourist areas the level of reliance on the industry is extremely high.

Tourism is the largest source of foreign exchange for the country. The revenue from tourism plays a significant part in helping central and local government fund economic and social policies. Special industry taxes have gone directly into social development, healthcare and education, all of which are often referred to as 'soft infrastructure'. However, tourism has also spurred the development of 'hard infrastructure' such as roads, telecommunications and airports. Also, as attitudes within the industry itself are changing, larger hotels and other aspects of the industry have become more socially conscious. Classic examples are the funding of local social projects.

National Parks and ecotourism

Figure 14 shows the location of Jamaica's three National Parks. A further six sites have been identified for future protection. The Jamaican government sees the designation of the parks as a positive environmental impact of tourism. Entry fees to the National Parks pay for conservation. The desire of tourists to visit these areas and the need to conserve the environment to attract future tourism drive the designation and management process.

The two marine parks are attempting to conserve the coral reef environments off the west coast of Jamaica. They are at risk from damage by overfishing, industrial pollution and mass tourism. The Jamaica Conservation and Development Trust is responsible for the management of the National Parks while the National Environmental Planning Agency has overseen the government's sustainable development strategy since 2001.

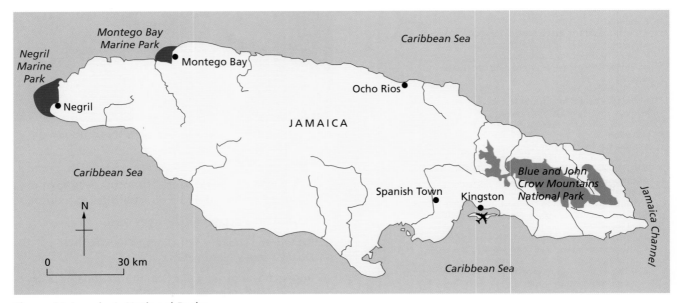

Figure 14 Jamaica's National Parks

Figure 15 A beach fringed with palm trees in Montego Bay Marine Park

Ecotourism is a developing sector of the industry with, for example, raft trips on the Rio Grande river increasing in popularity. Tourists are taken downstream in very small groups. The rafts, which rely solely on manpower, leave singly with a significant time gap between them to minimise any disturbance to the peace of the forest.

Community tourism

Considerable efforts are being made to promote **community tourism** so that more money filters down to the local population and small communities. The Sustainable Communities Foundation through Tourism (SCF) programme has been particularly active in central and south-west Jamaica. Community tourism is seen as an important aspect of '**pro-poor tourism**'. This is tourism that results in increased net benefits for poor people.

The Jamaica Tourist Board (JTB) is responsible for marketing the country abroad. Recently it used the fact that Jamaica was one of the host countries for the 2007 Cricket World Cup to good effect. The JTB also promotes the positive aspects of Jamaican culture and the Bob Marley Museum in Kingston has become a popular attraction. Such attractions are an important part of Jamaica's objective of reducing seasonality.

The disadvantages of tourism

The high or 'winter' season runs from mid-December to mid-April when hotel prices are highest. The rainy season extends from May to November. It has been estimated that 25 per cent of hotel workers are laid off during the off-season.

This has a major adverse impact on the standard of living of households reliant on the tourist industry. It also of course means that expensive tourism infrastructure is under-used for part of the year.

Although seasonality is seen as the major problem associated with tourism in Jamaica, other negative aspects include:

- the environmental impact of tourism which includes traffic congestion and pollution at popular locations
- the destruction of the natural environment to make way for tourism infrastructure
- the heavy use of resources, particularly water, by hotels
- under-use of facilities in the off-season
- socio-cultural problems illustrated by the behaviour of some tourists which clashes with the island's traditional morals (though some visitors also have a negative image of Jamaica because of its levels of violent crime and harassment).

Activities

1 Explain the importance of tourism to the economy of Jamaica.

2 Discuss the main problems associated with tourism.

■ Sustainable tourism

Sustainable tourism is tourism organised in such a way that its level can be sustained in the future without creating irreparable environmental, social and economic damage to the receiving area.

As the level of global tourism increases rapidly it is becoming more and more important for the industry to be responsibly planned, managed and monitored. Tourism operates in a world of finite resources where its impact is becoming of increasing concern to a growing number of people. At present, only 5 per cent of the world's population have ever travelled by plane but this is undoubtedly going to increase substantially.

Environmental groups are keen to make travellers aware of their '**destination footprint**'. This is the environmental impact caused by an individual tourist on holiday in a particular destination. Such groups are urging people to:

- 'fly less and stay longer'
- carbon-offset their flights
- consider 'slow travel'.

For the last, tourists consider the impact of their activities both for individual holidays and in the longer term. For example, they may decide that every second holiday will be in their own country (not using air transport). It could also involve using locally run guesthouses and small hotels as opposed to hotels run by international chains. This enables more money to remain in local communities.

Virtually every aspect of the industry now recognises that tourism must become more sustainable. **Ecotourism** is at the leading edge of this movement. This is a specialised form of tourism where people experience relatively untouched natural environments, such as coral reefs, tropical forests and remote mountain areas, and ensure that their presence does no further damage to these environments.

Protected areas

Over the course of the last 130 years or so, more and more of the world's most spectacular and ecologically sensitive areas have been designated for protection at various levels. The world's first National Park was established at Yellowstone in the USA in 1872. Now there are well over 1000 worldwide. Many countries have National Forests, Country Parks, Areas of Outstanding Natural Beauty, World Heritage Sites and other designated areas which merit special status and protection. Wilderness Areas with the greatest restrictions on access have the highest form of protection.

In many countries and regions there are often differences of opinion when the issue of special

protection is raised. For example, in some areas jobs in mining, forestry and tourism may depend on developing presently unspoilt areas. So it is not surprising that values and attitudes can differ considerably when big decisions about the future of environmentally sensitive areas are being made. Often, a clear distinction has to be made between the objectives of **preservation** and **conservation**. Preservation is maintaining a location exactly as it is and not allowing development. Conservation is allowing for developments that do not damage the character of a destination.

Tourist hubs

The concept of tourism hubs or clusters is a model that has been applied in a number of locations. The idea is to concentrate tourism and its impact in one particular area so that the majority of the region or country feels little of the negative impacts of the industry. Benidorm in Spain and Cancun in Mexico are examples where the model was adopted, but both locations show how difficult it is to confine tourism within preconceived boundaries as the number of visitors increases and people want to travel beyond tourist enclaves.

Quotas

Quotas seem to be one of the best remedies on offer. The UK Centre for Future Studies has suggested a lottery-based entrance system, an idea endorsed by Tourism Concern. Here, the number of visitors would not be allowed to exceed a sustainable level. This is an idea we are likely to hear much more about in the future.

Figure 16 Sand dune restoration works, County Kerry, Ireland

Figure 17 Combating severe informal footpath erosion on Mt Vesuvius, Italy

Case Study

Ecotourism in Ecuador

The country's tourism strategy has been to avoid becoming a mass market destination but to market 'quality' and 'exclusivity' instead, in as eco-friendly a way as possible.

Ecotourism has helped to bring needed income to some of the poorest parts of the country. It has provided local people with a new alternative way of making a living. As such it has reduced human pressure on ecologically sensitive areas.

The main geographical focus of ecotourism has been in the Amazon rainforest around Tena, which has become the main access point. The ecotourism schemes in the region are usually run by small groups of indigenous Quichua Indians (Figure 18).

Activities

1 What do you think of the idea of quotas for visitor numbers at certain locations?

2 What do you understand by the concept 'slow travel'?

3 Describe the operation of ecotourism in Ecuador's rainforest.

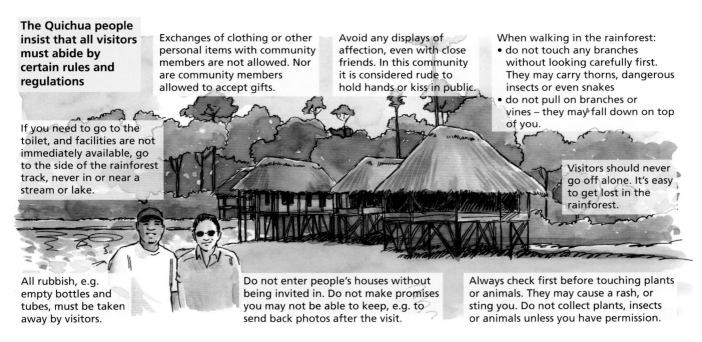

The Quichua people insist that all visitors must abide by certain rules and regulations

Exchanges of clothing or other personal items with community members are not allowed. Nor are community members allowed to accept gifts.

Avoid any displays of affection, even with close friends. In this community it is considered rude to hold hands or kiss in public.

When walking in the rainforest:
• do not touch any branches without looking carefully first. They may carry thorns, dangerous insects or even snakes
• do not pull on branches or vines – they may fall down on top of you.

If you need to go to the toilet, and facilities are not immediately available, go to the side of the rainforest track, never in or near a stream or lake.

Visitors should never go off alone. It's easy to get lost in the rainforest.

All rubbish, e.g. empty bottles and tubes, must be taken away by visitors.

Do not enter people's houses without being invited in. Do not make promises you may not be able to keep, e.g. to send back photos after the visit.

Always check first before touching plants or animals. They may cause a rash, or sting you. Do not collect plants, insects or animals unless you have permission.

Figure 18 Ecotourism in Ecuador's rainforest

Energy and water resources

Fuelwood in LEDCs

In developing countries about 2.5 billion people rely on fuelwood, charcoal and animal dung for cooking. Fuelwood and charcoal are collectively called fuelwood, which accounts for just over half of global wood production. Fuelwood provides much of the energy needs for sub-Saharan Africa. It is also the most important use of wood in Asia. Figure 1 shows the number of people living without electricity in the world.

Although at least one study claims that the global demand for fuelwood peaked in the mid-1990s, there can be no doubt that there are severe shortages in many countries. This is a major factor in limiting development.

In developing countries the concept of the 'energy ladder' is important. Here, a transition from fuelwood and animal dung to 'higher level' sources of energy occurs as part of the process of economic development. Income, regional electrification and

South Asia	706 million
Sub-Saharan Africa	547 million
East Asia	224 million
Other regions	101 million

Figure 1 People living without electricity, 2004

household size are the main factors influencing the demand for fuelwood. Thus, forest depletion is initially heavy near urban areas but slows down as cities become wealthier and change to other forms of energy. More isolated rural areas are the most likely to lack connection to an electricity grid. It is in such areas that the reliance on fuelwood is greatest. Wood is likely to remain the main source of fuel for the global poor in the foreseeable future.

The collection of fuelwood does not cause deforestation on the same scale as the clearance of land for agriculture, but it can seriously deplete wooded areas. The use of fuelwood is the main cause of indoor air pollution in developing countries. Indoor air pollution is responsible for 1.5 million deaths in developing countries every year. More than half of these deaths are of children below the age of 5.

Case Study

Fuelwood in Mali

Mali in West Africa is a huge landlocked and extremely poor country (Figure 2). It is in the Sahel, a region threatened by drought and desertification. The northern 65 per cent of the country is desert or semi-desert. Most people depend for their livelihood on the environment, by

Figure 2 Mali

farming, herding or fishing. The population of 12 million is growing at about 3 per cent a year.

Energy is a big issue in Mali. The country has no fossil fuel resources of its own. This means that fossil fuels have to be imported through neighbouring coastal countries, increasing costs considerably. For example, generating costs for grid electricity are twice as high as for Côte d'Ivoire. Imported petroleum accounts for 8 per cent of the country's trade balance. This is a major financial cost for a very poor country.

In rural areas 80 per cent of energy needs are supplied by firewood and charcoal. This uses over 50 million tonnes of national forest reserves every year. Kerosene lamps, torches and rechargeable car batteries are used for lighting. The latter also facilitate TV and radio. A small minority have a generator or solar panels.

Woodcutting is a rural industry in itself which provides the only source of employment for many people. Areas close to the main urban areas are particularly vulnerable to this activity. In 1994, 600,000 tonnes of wood were cut for use in the capital Bamako. By 2006 this had increased to almost 900,000 tonnes. The government predicts that if nothing is done to reverse this trend the demand for wood will be greater than supply by 2010.

Less than 12 per cent of the population has access to formal electricity (Figure 3). The contrast between urban and rural areas is huge. Most people connected to mains electricity live in the capital Bamako and the main towns. In rural areas less than 1 per cent of the population has access to mains electricity.

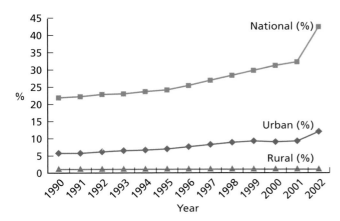

Figure 3 Electricity access in Mali, 1990–2000

Activities

1 **a)** What is fuelwood?
   b) Why is it such an important source of energy in the developing world?

2 Describe and explain the energy situation in Mali.

Non-renewable and renewable energy supplies

Non-renewable sources of energy are the **fossil fuels** and nuclear fuel. Fossil fuels consisting of hydrocarbons (coal, oil and natural gas) were formed by the decomposition of prehistoric organisms in past geological periods. These resources are finite so as they are used up the supply that remains is reduced. Eventually, these non-renewable resources could become completely exhausted.

Renewable energy can be used over and over again. These resources are mainly forces of nature that are **sustainable** and which usually cause little or no environmental pollution. Renewable energy includes hydro-electric, biogas, wind, solar, geothermal, tidal and wave power.

At present, non-renewable resources dominate global energy. The challenge is to transform the global **energy mix** to achieve a better balance between renewables and non-renewables.

There is a huge gap in energy consumption between rich and poor countries. Wealth is the main factor explaining the energy gap. The use of energy can improve the quality of life in so many ways. That is why most people who can afford to buy cars, televisions and washing machines do so. However, there are other influencing factors, with climate at the top of the list. Figure 4 shows the energy gap for high, middle and low-income countries.

The demand for energy has grown steadily over time. Figure 5 (page 142) shows a global increase

Kg of oil equivalent/person	
High-income countries	5435
Middle-income countries	1390
Low-income countries	494

Figure 4 Per capita energy consumption in low-, middle- and high-income countries, 2003

of over 60 per cent between 1981 and 2006. The fossil fuels dominate the global energy situation. Their relative contribution in 2006 was: oil 36 per cent, coal 28 per cent, natural gas 24 per cent. In contrast, hydro-electricity and nuclear energy accounted for about 6 per cent each. Figure 5 shows commercially traded fuels only. It excludes fuels such as wood, peat and animal waste which, though important in many countries, are unreliably documented in terms of consumption statistics.

Consumption by type of fuel varies widely by world region (Figure 7):

- **Oil** Nowhere is the contribution of oil less than 30 per cent and it is the main source of energy in four of the six regions shown in Figure 7. In the Middle East it accounts for approximately 50 per cent of consumption.
- **Coal** Only in the Asia Pacific region is coal the main source of energy. In contrast it accounts for less than 5 per cent of consumption in the Middle East and South and Central America.
- **Natural gas** Natural gas is the main source of energy in Europe and Eurasia and it is a close second to oil in the Middle East. Its lowest share of the energy mix is 11 per cent in Asia Pacific.

Figure 6 Aircraft refuelling at Gatwick airport, UK

- **Hydro-electricity (HEP)** The relative importance of HEP is greatest in South and Central America (28 per cent). Elsewhere its contribution varies from 6 per cent in Africa to less than 1 per cent in the Middle East.
- **Nuclear energy** Nuclear energy is not presently available in the Middle East and it makes the smallest contribution of the five energy sources in Asia Pacific, Africa and South and Central America. It is most important in Europe and Eurasia and North America.

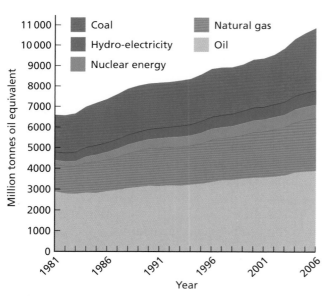

World primary energy consumption grew more slowly in 2006 but growth remained just above the 10-year average. Oil was the slowest-growing fuel, while coal was the fastest-growing. Although oil remains the world's leading energy source, it has lost market share to coal and natural gas in the past decade.

Figure 5 Changes in world energy consumption by type, 1981–2006

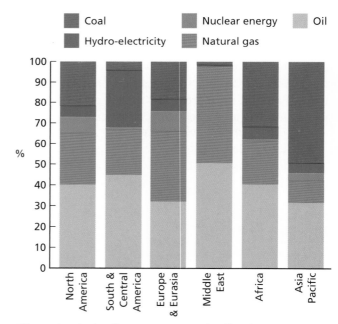

Oil remains the leading energy source in all regions except Asia Pacific and Europe & Eurasia. Coal dominates in the Asia Pacific region, while natural gas is the leading fuel in Europe & Eurasia. The Asia Pacific region accounted for two-thirds of global energy consumption growth in 2006.

Figure 7 Regional energy consumption patterns, 2006

Figure 8 Fuel station on the River Amazon

Renewable energy supplies
Hydro-electricity

Of the traditional five major sources of energy, HEP is the only one that is renewable. It is by far the most important source of renewable energy. The 'big four' HEP nations of China, Canada, Brazil and the USA account for over 46 per cent of the global total (Figure 9).

Most of the best HEP locations are already in use so the scope for more large-scale development is limited. However, in many countries there is scope for small-scale HEP plants to supply local communities.

Although HEP is generally seen as a clean form of energy, it is not without its problems which include:

- large dams and power plants that can have a huge negative visual impact on the environment
- the obstruction of the river for aquatic life
- deterioration in water quality
- large areas of land that may need to be flooded to form the reservoir behind the dam
- submerging large forests without prior clearance that can release significant quantities of methane, a greenhouse gas.

	Country	Million tonnes oil equivalent	% share of world total
1	China	94.3	13.7
2	Canada	79.3	11.5
3	Brazil	79.2	11.5
4	USA	65.9	9.6
5	Russia	39.6	5.8
6	Norway	27.1	3.9
7	India	25.4	3.7
8	Japan	21.5	3.1

Figure 9 HEP consumption, 2006

New alternative energy sources

The first major wave of interest in new alternative energy sources resulted from the energy crisis of the early 1970s. Then the relatively low price of oil in the 1980s, 1990s and the opening years of the present century dampened down interest in these energy sources. However, renewed concerns about energy in recent years and corresponding price increases have kick-started the alternative energy industry again.

The main drawback to the new alternative energy sources is that they invariably produce higher-cost electricity than traditional sources. The cost gap with non-renewable energy is narrowing, though.

Solar power

There are two systems for producing solar electricity at present:

- Photovoltaic systems – these are solar panels that convert sunlight directly into electricity. By the end of 2002, 1500 MW had been installed globally. This is only the size of one large coal-fired power station. The leading countries were Japan (627 MW), Germany (295 MW) and the USA (212 MW).
- Thermal power plants – total global installed capacity at the end of 2002 was 364 MW, most in the form of nine power plants in the Mohave Desert in southern California. Located on three sites, the plants vary in size from 14 to 80 MW. Parabolic trough technology is used to collect the sun's rays. Steam is generated at 400 °C to drive the turbines which produce 345 MW. This is enough to meet the demand from more than half a million people.

Figure 10 Solar power

Wind power

Global wind-generated electricity capacity totalled 47,300 MW at the end of 2004. Figure 11 shows that almost 67 per cent of global wind power is concentrated in just three countries, with Germany leading the way. Only seven countries produced more than 1000 MW in 2004.

Apart from establishing new wind energy sites, **repowering** could also play an important role. This means replacing old wind turbines with new engines which give a better performance.

Biomass

Biomass is organic matter from which energy can be produced. The direct domestic use of biomass, for example burning firewood, is a major source of energy in many developing countries. But in terms of the future it is biomass schemes that a) generate electricity and b) act as a substitute for oil that are the centre of attention. Figure 13 shows production in the leading biofuel countries in the world.

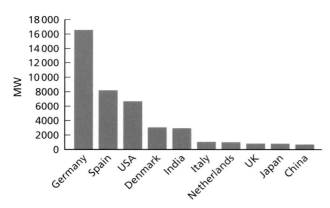

Figure 11 Global wind power 'top 10'

Figure 12 Wind farm in California

Due to the high price of oil, biofuels are increasing in popularity as sources of power for trucks, cars and planes. In Brazil, all fuel sold to motorists is 22–26 per cent ethanol. Brazil is the world's largest producer of ethanol, which is distilled from sugar cane. However, a new generation of cars (flex-fuel vehicles) can run on ethanol alone, which is about half the price of the alternative at the petrol pumps. Brazil exports ethanol to countries such as Japan and South Korea as they start to diversify energy consumption away from oil. The USA is just behind Brazil in ethanol production, with corn being the crop source. In Germany, 'biodiesel' (from rapeseed) is increasing in production. In over 30 countries a variety of crops are now being grown for fuel. With rising grain prices, food and fuels may be in competition in a number of countries.

Geothermal power

Geothermal energy is the natural heat found in the Earth's crust in the form of steam, hot water and hot rock. Rainwater may percolate several kilometres in permeable rocks where it is heated due to the Earth's geothermal gradient. This is the rate at which temperature rises as depth below the surface increases. The average rise in temperature is about 30 °C per km, but the gradient can reach 80 °C near plate boundaries.

This source of energy can be used to produce electricity or its hot water can be used directly for industry, agriculture, bathing and cleansing. For example, in Iceland, hot springs supply water at 86 °C to 95 per cent of the buildings in and around Reykjavik. At present virtually all the geothermal power plants in the world operate on steam resources (Figure 14), having an extremely low environmental impact.

By 2005 the global capacity of geothermal electricity had reached 8900 MW, with considerable new development underway. The number of

Million tonnes oil equivalent						
	USA	Brazil	EU	India	China	World
Ethanol	7.50	8.17	0.48	0.15	0.51	17.07
Biodiesel	0.22	0.05	2.53	0	0	2.91
Total	7.72	8.22	3.01	0.15	0.51	19.98

Figure 13 Leading biofuel countries, 2005

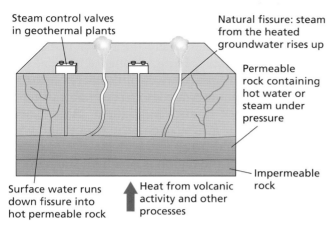

Figure 14 Geothermal power

Figure 15 Geothermal power plant in Iceland

countries producing power from this source could rise from 21 in 2000 to 46 in 2010, with global capacity reaching 13,500 MW.

The USA is the world leader in geothermal electricity, with plants in Alaska, California, Hawaii, Nevada and Utah. Total production accounts for 0.37 per cent of the electricity used in the USA. Other leading geothermal electricity countries are the Philippines, Italy, Mexico, Indonesia, Japan, New Zealand and Iceland.

Activities

1 Explain the difference between renewable and non-renewable sources of energy.

2 Describe and attempt to explain the variation in energy mix by world region.

3 Apart from hydro-electricity, why does renewable energy contribute so little to global energy supply?

4 For the country in which you live, find out which forms of renewable energy are used and how much they contribute to total energy production.

Factors affecting the siting of power stations

Hydro-electric power stations

The location of HEP plants is determined to a great extent by relief and drainage. An ideal location would have:

- a major fast-flowing river
- steep-sided valley cut into hard impermeable rock
- few people living in the area to be flooded by the reservoir
- existing transport networks close at hand
- easy access to existing electricity transmission corridors.

Few sites combine all these factors and it is the physical attributes that are the most important. When a possible site is being assessed, difficult decisions concerning costs have to be made.

The world's second largest HEP scheme (after the Three Gorges project in China) is located on the Paraná River in South America (Figure 16). It is a joint venture between Brazil and Paraguay. The Paraná is the second longest river in South America, with a large, reliable flow of water. The annual average discharge is 17,293 m^3 per second. The hard, impermeable rock was ideal for constructing both the

Figure 16 Location of Itaipù

dam and the reservoir. The depth of the valley and the relief of the wider area flooded for the reservoir means that Itaipù has the lowest flooded area per unit of power production of all the major HEP schemes in Brazil. However, about 40,000 people had to be moved from their homes to make way for it. With the towns of Foz do Iguaçu and Puerto Strossner located nearby, a reasonable level of infrastructure was already in place. Itaipù has an installed capacity of 14 GW, with 20 generating units of 700 MW each. In 2005 it supplied 93 per cent of Paraguay's electricity and 20 per cent of Brazil's.

Nuclear power stations

The countries that produce nuclear electricity are generally those that have had concerns about providing enough energy from other sources. This is because a) nuclear energy is the most controversial source of power and b) it is often more costly than some other sources of electricity.

In 2007, China had 11 nuclear reactors spread over four locations. Four of the reactors are at Daya Bay, Guangdong in south-east China, two at the Daya Bay nuclear power plant and two at the Ling Ao nuclear power plant.

A coastal location permits seawater to be used in the cooling process. Nuclear power plants require large quantities of water for this purpose. The hard rock of the area provides a solid foundation for these large and heavy installations. There is also no major threat from earthquakes or faulting in the area. The major cities where the power is used are not too far away (Hong Kong about 50 km, Shenzhen 40 km), so relatively little energy is lost in transmission, but they are also a reasonable distance away in case of a nuclear accident. A sufficient supply of labour is available in the region and with major cities not too far away the general infrastructure of the region is of a high standard.

China has ambitious plans to expand its nuclear electricity capacity and may eventually become the most important country in the world in this respect.

Thermal power stations

The characteristics of thermal power stations are that they:

- burn fossil fuels to create steam to drive the turbines
- are located beside a major river, lake or by the sea to provide the large quantities of water to cool the steam that drives the turbines
- require the infrastructure (port, rail terminal, pipeline) to receive and store the significant quantities of fossil fuel (coal and oil) needed to supply the power station
- provide the balance between avoiding large concentrations of people but are not too remote from consumers to minimise the losses of power along transmission corridors.

Kingsnorth is a major 2000 MW thermal power station in south-east England. It is located on the Hoo Peninsula on the banks of the Medway estuary where there is access to ample quantities of water. It is a dual-fired power station – each of its four main

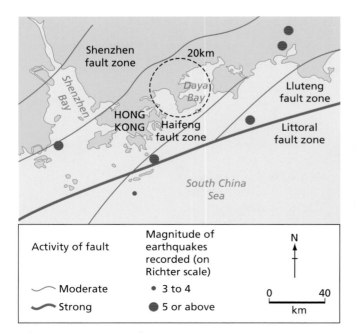

Figure 17 Location of Daya Bay

Figure 18 Aerial view of Kingsnorth power station

units is capable of using both coal and oil. Kingsnorth has a port facility to allow the importation of these fuels. Figure 18 shows the large area of land required to store coal and oil. The power station is adjacent to farmland and there are no significant residential areas nearby. However, it is not a great distance from the markets it serves.

Activities

1 Suggest why Itaipù was chosen as the location for Brazil's largest hydro-electric power station.

2 Explain the important location factors for a nuclear power station.

3 Describe and account for the location of Kingsnorth thermal power station.

■ Water uses, competition and management

Global pressures on water supply

As population increases and living standards rise in many countries, the demand for water for agriculture, industry and domestic use continues to grow. In some regions the demand far exceeds the supply. Here, water shortages have a severe impact on local people and on the potential for development. This results in competition for the

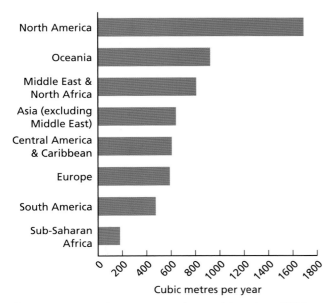

Figure 19 Per capita water use by world region, 2000

Figure 20 Dried-up river bed in northern Spain

available water resources and the need for careful management. Figure 19 shows per capita water use by world region.

The situation becomes more complicated if the resource, such as water from a major river, is shared by more than one country. An example is the River Nile. The Nile Water agreement of 1929 gave Egypt the largest individual share of Nile water, but countries upstream of Egypt such as Tanzania and Kenya, are becoming increasingly unhappy about this. When Kenya announced its decision to withdraw from the 1929 agreement, the Egyptian water minister described it as 'an act of war'.

Case Study

The water problem in the south-western USA

The USA is a huge consumer of water. Over the country as a whole there would not seem to be a water problem. However, the western states of the USA, covering 60 per cent of the land area with 40 per cent of the total population, receive only 25 per cent of the country's mean annual precipitation. Yet each day the west uses as much water as the east.

The west has prospered due to a huge investment in water transfer schemes. This has benefited agriculture, industry and settlement. Hundreds of aqueducts take water from areas of surplus to areas of shortage. The federal government has paid most of the bill but now the demand for water is greater than the supply. If the west is to continue to expand, a solution to the water problem must be found.

Figure 21 Desert region in south-west USA

Although much of the west is desert or semi-desert, large areas of dry land have been transformed into fertile farms and sprawling cities. It all began with the Reclamation Act of 1902 which allowed the building of canals, dams and HEP systems in the states that lie, all or in part, west of the 100th meridian. Water supply was to be the key to economic development in general, benefiting not only the west but the USA as a whole.

California has benefited most from this investment in water supply. There is a great imbalance between the distributions of precipitation and population in the state. Seventy per cent of runoff originates in the northern one-third of the state but 80 per cent of the demand for water is in the southern two-thirds. While irrigation is the prime water user, the sprawling urban areas have also greatly increased demand. The 3.5 million hectares of irrigated land in California are situated mainly in the Imperial, Coachella, San Joaquin and the lower Sacramento valleys. Figure 22 shows the major component parts of water transfer and storage in the state.

Agriculture uses more than 80 per cent of the state's water, though it accounts for less than a tenth of the economy. Water development, largely

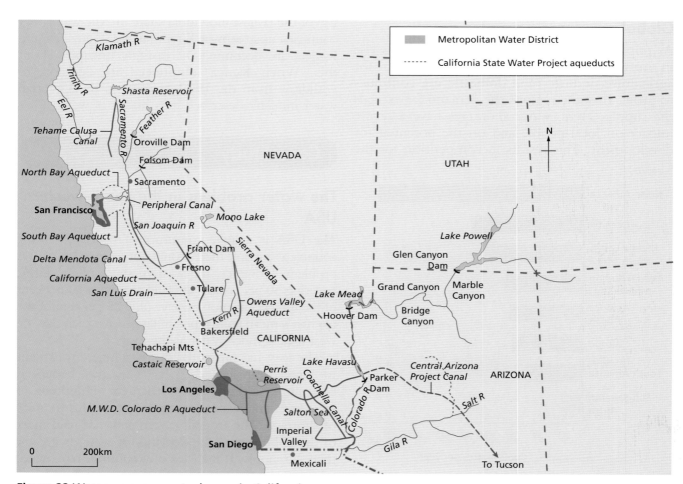

Figure 22 Water management schemes in California

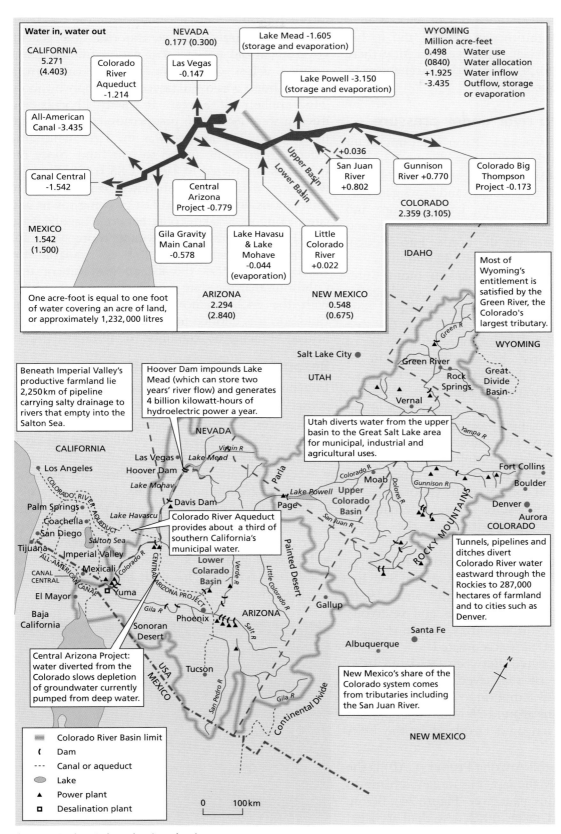

Water in, water out

CALIFORNIA
5.271
(4.403)

NEVADA
0.177 (0.300)

Lake Mead -1.605
(storage and evaporation)

WYOMING
Million acre-feet
0.498   Water use
(0840)   Water allocation
+1.925   Water inflow
-3.435   Outflow, storage
           or evaporation

Colorado
River
Aqueduct
-1.214

Las Vegas
-0.147

Lake Powell -3.150
(storage and evaporation)

All-American
Canal -3.435

+0.036

San Juan
River
+0.802

Gunnison
River +0.770

Colorado Big
Thompson
Project -0.173

Canal Central
-1.542

Central
Arizona
Project -0.779

COLORADO
2.359 (3.105)

MEXICO
1.542
(1.500)

Gila Gravity
Main Canal
-0.578

Lake Havasu
& Lake
Mohave
-0.044
(evaporation)

Little
Colorado
River
+0.022

IDAHO

Most of
Wyoming's
entitlement is
satisfied by the
Green River, the
Colorado's
largest tributary.

One acre-foot is equal to one foot
of water covering an acre of land,
or approximately 1,232,000 litres

ARIZONA
2.294
(2.840)

NEW MEXICO
0.548
(0.675)

Beneath Imperial Valley's
productive farmland lie
2,250km of pipeline
carrying salty drainage to
rivers that empty into the
Salton Sea.

Hoover Dam impounds Lake
Mead (which can store two
years' river flow) and generates
4 billion kilowatt-hours of
hydroelectric power a year.

Utah diverts water from the upper
basin to the Great Salt Lake area
for municipal, industrial and
agricultural uses.

Colorado River Aqueduct
provides about a third of
southern California's
municipal water.

Tunnels, pipelines and
ditches divert
Colorado River water
eastward through the
Rockies to 287,000
hectares of farmland
and to cities such as
Denver.

Central Arizona Project:
water diverted from the
Colorado slows depletion
of groundwater currently
pumped from deep water.

New Mexico's share of the
Colorado system comes
from tributaries including
the San Juan River.

Legend:
- Colorado River Basin limit
- Dam
- Canal or aqueduct
- Lake
- Power plant
- Desalination plant

0   100km

Figure 23 The Colorado river basin

financed by the federal government, has been a huge subsidy to California in general and to big water-users in particular. However, recently there has been a move to bring the price mechanism to bear on water resources.

The Colorado: a river under pressure

The 2333 km long Colorado River is an important source of water in the south-west (Figure 23, page 149). The river rises 4250 m up in the Rocky Mountains of northern Colorado and flows generally south-west through Colorado, Utah, Arizona and between Nevada and Arizona, and Arizona and California before crossing the border into Mexico. The river drains an area of about 632,000 km².

In 1912 Joseph Lippincott, seeking future water supplies for the growing city of Los Angeles, described the Colorado as 'an American Nile awaiting regulation'. The Colorado was the first river system in which the concept of multiple use of water was attempted by the US Bureau of Reclamation. In 1922 the Colorado River Compact divided the seven states of the basin into two groups: Upper Basin and Lower Basin. Each group was allocated 9.25 trillion litres of water annually, while a 1944 treaty guaranteed a further 1.85 trillion litres to Mexico. Completed in 1936, the Hoover Dam and Lake Mead marked the beginning of the era of artificial control of the Colorado.

Despite the inter-state and international agreements, major problems over the river's resources have arisen:

- Although the river was committed to deliver 20.35 trillion litres every year, its annual flow has averaged only 17.25 trillion litres since 1930. Evaporation from artificial lakes and reservoirs has removed 2.45 trillion litres, and in drought periods this shortfall is accentuated.
- Demand has escalated. Between 1970 and 1990 the population of the seven Compact states increased from 22.8 million to 36.1 million. The river now sustains around 25 million people and 820,000 ha of irrigated farmland in the USA and Mexico.

The $4 billion Central Arizona Project (CAP) is the latest, and probably the last, big money scheme to divert water from this great river (Figure 24). Before CAP, Arizona had taken much less than its legal entitlement from the Colorado; it could not afford to build a water transfer system from the Colorado to its main cities and at the time the federal government did not feel that national funding was justified. Most of the state's water came from aquifers but it was overdrawing this supply by about 2467 million m³ a year. If thirsty Phoenix and Tucson were to remain prosperous, something had to be done. The answer was the CAP which the federal government agreed to part fund. Since the CAP was completed in 1992, 1.85 trillion litres of water a year has been distributed to farms, Indian reservations, industries and fast-growing towns and cities along its 570 km route between Lake Havasu and Tucson. However, providing more water for Arizona has meant that less is available for California. In 1997 the federal government told California that the state would have to learn to live with the 5427 million m³ of water from the Colorado it is entitled to under the 1922 Compact, instead of taking 6416 million m³ a year.

Resource management strategies

Implementation of the following strategies would conserve considerable quantities of water:

- Measures to reduce leakage and evaporation losses. Up to 25 per cent of all water moved is currently lost in these ways.
- Recycling water in industry where, for example, it takes 225,000 litres to make 1 tonne of steel.
- Recycling municipal sewage for watering lawns, gardens and golf courses could be implemented or extended, as Los Angeles has already shown.
- Introducing more efficient toilet systems which use only 6.5 litres for each flush instead of the conventional 26 litres.

Figure 24 Part of the Central Arizona Project

- Charging more realistic prices for irrigation water. Many farmers pay only one-tenth of the true cost of water pumped to them; the rest is subsidised by the federal government. When long-term water contracts are eventually renewed, prices could be raised to more economic levels.
- Extending the use of drip irrigation systems which allocate specific quantities of water to individual plants, and which are 100 times more efficient than the open-ditch system still used by many farmers; or sprinkler systems, which are up to ten times more efficient than open-ditch irrigation.
- Changing from highly water-dependent crops such as rice and alfalfa to those needing less water.
- Changing the law to permit farmers to sell surplus water to the highest bidders. Since 1992, this has been allowed in California, where an emerging network of specialist brokers sells 'agricultural water' to cities for less than they already pay but at a profit for the farmers.
- Requiring both cities and rural areas to identify the source of water to be used before new developments can commence. This proposal, first mooted in southern California in 1994, proved to be politically unacceptable.

Future options

- Developing new groundwater resources. Although groundwater has been heavily depleted in many areas, in regions of water surplus such as northern California they remain virtually untapped. However, the transfer of even more water from such areas would probably prove politically unacceptable.
- It has been claimed that various techniques of weather modification, especially cloud seeding, can provide water at reasonable cost. However, environmental and political considerations cannot be ignored here.
- In 1991, after several years of drought, the city of Santa Barbara approved the construction of a $37.4 million desalination plant. Although much too expensive for irrigation water, it is likely that more will be built for urban use.
- Exploiting the frozen reserves of Antarctic water. Serious proposals have been made to find a 100 million tonne iceberg (1.5 km long, 300 m wide, 270 m deep) off Antarctica, wrap it in sailcloth or thick plastic, and tow it to southern California. The critical questions here are cost, evaporation loss and the environmental effects of anchoring such a huge block of ice off an arid coast.
- There is now general agreement that planning for the future water supply of the south-west should embrace all practicable options. Sensible management of this vital resource should not rule out any feasible strategy if this important region is to sustain its economic viability and growing population.

Activities

1 Describe the imbalance in population and precipitation between the eastern and western parts of the USA.

2 Discuss the main uses of water in California.

3 Why is the Colorado River under so much pressure?

4 Explain the resource management strategies that can be used to try to improve the balance between supply and demand.

Environmental risks and benefits: resource conservation and management

■ Sustainable development and resource management/conservation in different environments

As the scale of global economic activity has increased, bringing considerable benefits to many people, the strain on the natural environment has become more obvious. Every aspect of human activity has an impact on the environment. The case studies that follow in this section examine some of the risks and benefits associated with agriculture, manufacturing, energy, tourism and transport.

The two key terms that have become increasingly important in terms of economic activity are resource management and sustainable development. **Resource management** is the control of the exploitation and use of resources in relation to environmental and economic costs. **Sustainable development** is a carefully calculated system of resource management which ensures that the current level of exploitation does not compromise the ability of future generations to meet their own needs.

Resource management in the European Union

Figure 1 shows what has happened in so many of the world's fishing grounds. Without careful resource management fish stocks could be totally depleted in some areas. Yet it is often difficult to get countries to agree on what to do. The European Union's Common Fisheries Policy is perhaps the most advanced international attempt to manage the fishing grounds belonging to this group of countries. While the fishing industry in the EU frequently complains that the amount of fish it is allowed to catch (the total allowable catch) is too low, environmental groups argue that the total allowable catch is much too high and that fishing in EU waters cannot be sustainable in the long term. Other people have an interest too. For example, consumers worry that if less fish are caught the price will increase.

The European Union also tries to manage agriculture in its member countries through its Common Agricultural Policy (CAP). In the early years the CAP's generous incentives for farmers encouraged high levels of production and the farming of marginal lands. It didn't seem that much thought was given to the environmental and other adverse consequences of maximising production. However, as the disadvantages became more obvious the CAP was reformed to take greater account of the environment. For example, farmers can now receive payments for taking their land out of agricultural production (set-aside).

It can be argued that the CAP is still a long way from truly sustainable agriculture, but there is no doubt it is moving in the right direction. The development of sustainable policies often occurs in stages.

Environmental impact statements and pollution control

Most countries now require some form of environmental impact statement for major projects such as a new road, an airport or a large factory. The objective is to identify all the environmental

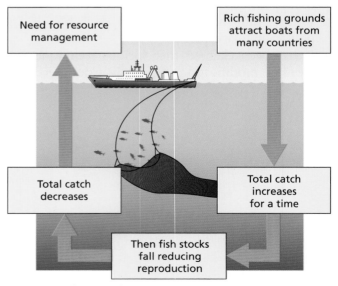

Figure 1 Fishing and resource management

consequences and to try to minimise these as far as possible.

Industry has spent increasing amounts on research and development to reduce pollution – the so-called 'greening of industry'. In general, after a certain stage of economic development the level of pollution will decline (Figure 2). This is because countries have become more aware of their environmental problems with higher levels of economic activity and they have also created the wealth to invest in improving the environment. The 1990s witnessed the first signs of 'product stewardship'. This is a system of environmental responsibility whereby producers take back a product, recycling it as far as possible, after the customer has finished with it. For example, in Germany the 1990 'take-back' law required car manufacturers to take responsibility for their vehicles at the end of their useful lives.

International action

Increasingly, successful policies developed in one country are being followed elsewhere. A good example is the role of ecotourism in rainforest conservation (Figure 3). International organisations are the only hope of getting to grips with the really big problems, such as climate change. The success of international co-operation in tackling the hole in the ozone layer gives us reasonable hope for the future.

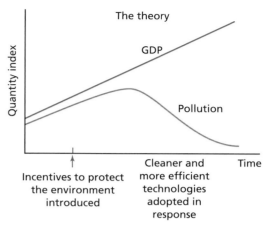

Figure 2 The relationship between GDP and pollution

Activities

1 What is meant by the terms **a)** *resource management* and **b)** *sustainable development*?

2 Explain the sequence of events shown in Figure 1.

3 Look at Figure 3. Describe and explain how ecotourism can enhance rainforest conservation.

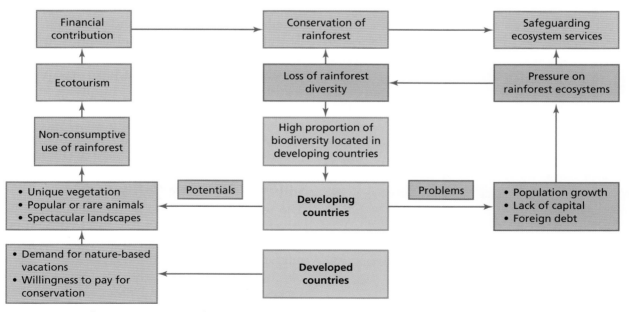

Figure 3 The role of ecotourism in rainforest conservation

Agriculture: risks and benefits

Case Study

Agricultural change in Argentina's Pampas

Traditionally, cattle rearing has dominated farming in the Pampas of Argentina. The Pampas is one of the world's great grasslands. It is a flat prairie with deep, fertile topsoil. The landscape has been dotted with 'estancias', large ranches where extensive commercial cattle rearing is practised. The estancias and the gauchos (cowboys) who work them are an important part of the country's identity and tradition.

Figure 5 The Pampas – cropland

However, rapid change is underway as crop production replaces cattle rearing over significant areas of the Pampas. According to the Argentine Rural Society, 10 million hectares of the Pampas have been ploughed up in the last 15 years. This is an area roughly two and a half times the size of Switzerland. There are undoubted benefits in this process as farmers are responding to changing patterns of global demand. But there are also risks involved in such a considerable change in land use.

Cropland now more profitable

More profitable soybeans and corn are replacing cattle on the grassy plains of the central Pampas. This change in agricultural land use has allowed Argentina's oilseed and grain industry to increase almost 50 per cent between 2003 and 2006. Global demand, particularly from China, has pushed up world prices considerably. In 2006, Argentina exported $9 billion of soybeans and soybean products, amounting to almost a fifth of the country's total exports. Argentina is the world's third largest soybean exporter after the USA and Brazil.

Cattle are moved to the north

The world-famous cattle of the Pampas are being driven to the harsher swamps and scrubland in the north of Argentina. This is a major agricultural migration. In these northern landscapes cattle have to contend with drought, flooding, poisonous snakes, vampire bats and piranhas. In 2007 it is estimated that 40,000 cattle died either through

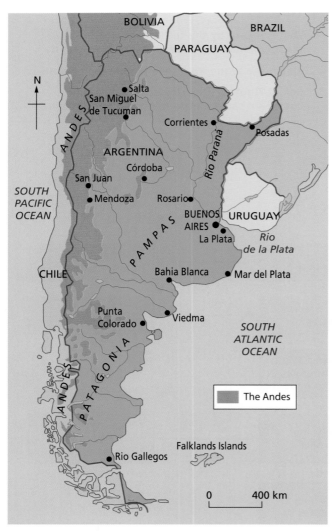

Figure 4 The Argentinian Pampas and the north

starvation or because of infected wounds from piranha bites. Temperatures can reach 40 °C in the far north. The north of Argentina now contains more than a third of the country's cattle, compared with less than 10 per cent in 2002.

The high reputation of Argentina's beef has been based on Aberdeen Angus and Hereford cattle. But now cattle owners are cross-breeding these cattle with Brahman strains from Brazil and India so that the new herds can cope better with the heat and poorer pastures of the north. Cross-breeding results in the flesh being less tender. Critics argue that this will reduce the quality and reputation of Argentina's beef, which was once considered by many to be 'the best beef in the world'. Argentina has the highest consumption of beef per capita in the world.

Environmental concerns

The change from pastoral to arable farming has considerably increased chemical input onto the land. This is having a significant impact on the ecosystem. The World Wide Fund for Nature (WWF) is concerned that the Pampas is now being over-farmed. This is endangering wildlife including South American ostriches, pumas and wildcats. The WWF is also concerned about the widespread destruction of native grasses.

Biofuels

An increasing amount of crop production in the Pampas is being used for biofuels. As in other large agricultural areas where this is happening, the practice is controversial because food prices have risen so significantly in recent years. The 'food v.

fuel' debate has a long way to run. Initially, biofuels were seen by many as a 'green' form of energy. Now, though, such a view is being strongly challenged. If global food prices continue to rise and biofuel production is recognised as a major factor, then governments may decide to restrict the amount of cropland devoted to biofuels.

Activities

1 Describe the location of the Pampas.

2 Suggest why the Pampas developed into an important cattle-rearing region.

3 Explain why so much of the grassland in the Pampas has been turned over to crop production.

4 Why is an increasing amount of crop production in the Pampas being used for biofuel production?

5 Why has the production of biofuels become so controversial in recent years?

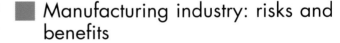

Manufacturing industry: risks and benefits

Case Study

China's Pearl River delta

The Chinese economy has attained such a size and is continuing to grow so rapidly that it is now being called 'the new workshop of the world', a phrase first applied to Britain during the height of its Industrial Revolution in the nineteenth century. However, in the main industrial areas the environment has been put under a huge strain, leaving China with some of the worst pollution problems on the planet. One of China's main industrial regions is the Pearl River delta. It faces the challenge of continuing to grow economically while trying to protect its environment.

The Pearl River delta region, an area the size of Belgium in south-east China (Figure 7, page 156), is the focal point of a massive wave of foreign investment into China. The Pearl River drains into

Figure 6 The North – new cattle lands

the South China Sea. Hong Kong is located at the eastern extent of the delta, with Macau situated at the western entrance. The region's manufacturing industries already employ 30 million people (Figure 8) but this will undoubtedly increase in the future.

The region has become the focus of such a high level of investment because of:

- the Cantonese work ethic that was so important in the rapid development of Hong Kong
- the very low cost of labour compared with alternative international locations
- the lack of unions
- an improving level of hard and soft infrastructure
- the proximity of suppliers – the efficiency of the supply chain now rivals the low cost of labour as the major location factor for some companies
- a welcoming regulatory environment.

Figure 7 The Pearl River delta

Figure 9 Industrial scene in the Pearl River delta

Shunde: the largest centre for the production of microwave ovens in the world. 40% of global production comes from just one huge factory (Galanz) in Shunde. Galanz exported 70% of the 15 million microwave ovens it made in 2002.
Shenzhen: The special economic zone estimates that it produces 70% of the world's photocopiers and 80% of its artificial Christmas trees. In 2002, the port of Shenzhen overtook both Rotterdam and Los Angeles to become the world's sixth largest container terminal. Shenzhen's huge shopping malls attract large numbers of wealthy residents from Hong Kong. The city attracts huge corporate buyers too. It is the global purchasing centre for Kingfisher and Wal-Mart, which together sourced $10 billion of goods from China in 2002.   In 2002 it was announced that seven new towns each capable of supporting 500,000 people are to be built outside Shenzhen special economic zone in the next decade.
Dongguan: specialises in running shoes, with 80,000 people employed in a single factory. The population of migrant workers is higher here than in any other Chinese city.
Zhongshan: the major world centre of the electric lighting industry.
Zhuhai: a major manufacturer of computer games, consoles and golf clubs. Land is being reclaimed from the South China Sea to facilitate further industrial expansion.
Guangzhou: the site of a large export-only Honda car plant. Nearly 70 of the top 500 transnational corporations are represented in the Guangzhou Development Zone. Industries include pharmaceuticals, specialist steel, cars, food, beverages, chemicals, electronics, electrical appliances.

Figure 8 Economic hotspots in the Pearl River Delta

Environmental problems

The three major environmental problems in the Pearl River delta are air pollution, water pollution and deforestation. In 2007, eight out of every ten rainfalls in Guangzhou were classified as acid rain. The high concentration of factories and power stations is the source of this problem, along with the growing number of cars in the province. The city has the worst acid rain problem in the province of Guangdong. The province's environmental protection bureau has reported that two-thirds of Guangdong's 21 cities were affected by acid rain in 2007. Overall, 45 per cent of the province's rainfall in 2007 was classified as acid rain.

Half of the wastewater in Guangdong's urban areas is not treated before it is dumped into rivers, compared with the national average of 40 per cent. Chemical oxygen demand (COD) is a key measurement of water pollution. Guangdong's government has pledged to reduce COD by 15 per cent by 2010 (from 2005 levels). It also aims to cut sulphur dioxide emissions by 15 per cent.

Almost all the urban areas have over-exploited their neighbouring uplands, causing a considerable reduction in vegetation cover. This has resulted in serious erosion.

The environmental protection bureau classifies the environmental situation as 'severe' and says the government is committed to take the 'necessary measures' to reduce pollution. Among the measures used to tackle the problems are a) higher sewage treatment charges b) stricter pollution regulations on factories and c) tougher national regulations on vehicle emissions.

Activities

1 Draw an annotated sketch map to show the main industrial areas in the Pearl River delta.

2 a) List the main environmental problems in the region.
   b) Discuss the causes of these problems.
   c) What is government trying to do to reduce these problems?

Case Study

■ Energy: risks and benefits

The Niger delta

Nigeria's oil is located in the Niger delta (Figure 10). Oil was first extracted here in 1956. The low-sulphur oil quickly gained important export markets around the world and by the mid-1970s Nigeria had joined OPEC, the influential Organisation of Petroleum Exporting Countries. Nigeria is the world's sixth largest oil exporter. Forty per cent of its production goes to the USA and 21 per cent to western Europe.

Production is dominated by five major transnational corporations (TNCs) – Shell, Total, Agip, ExxonMobil and Chevron. Together they operate 159 oilfields, 275 flow stations and 7250 km of pipelines. Many of the new oilfields are offshore.

In 1960, farm products such as palm oil and cacao beans accounted for nearly all of Nigeria's exports. Today, oil makes up 90 per cent of export earnings and 80 per cent of its revenue. The most populous country in Africa with 130 million people has gone from being self-sufficient in food to importing more than it produces. This is because both the government and many communities have neglected agriculture in pursuit of oil wealth.

Although Nigeria produces a large quantity of oil, its own refineries are old and poorly run, resulting in frequent breakdown. Thus the country also imports the bulk of its fuel. A recent United Nations report put the quality of life in Nigeria

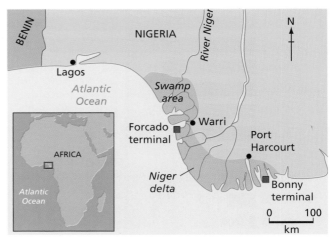

Figure 10 The Niger Delta

below all the other major oil-producing countries. It has been estimated that corruption siphons off as much as 70 per cent of annual oil revenues. Most Nigerians live on less than a dollar a day. Fifty years of oil extraction has failed to improve the lives of the majority of people. Critics blame the government and the transnational oil companies for this terrible state of affairs.

Nigeria nationalised its oil industry in 1971. In a joint venture arrangement, the Nigerian National Petroleum Corporation (owned by the government), owns 55–60 per cent of TNC oil operations. Each one of Nigeria's 36 state governments receives a share of oil money but 'trickle down' to the country's poor is extremely limited.

The Niger delta is one of the world's largest wetlands and Africa's largest remaining mangrove forest. It has suffered an environmental disaster from the extraction of oil:

- Oil spills, acid rain from gas flares and the stripping away of mangroves for pipeline routes have killed off fish.
- Between 1986 and 2003, more than 20,000 hectares of mangroves disappeared from the coast, mainly due to land clearing and canal dredging for oil and gas exploration.
- The oilfields contain large amounts of natural gas. This is generally burnt off as flares rather than being stored or reinjected into the ground. Hundreds of flares have burned continuously for decades. This causes acid rain and releases greenhouse gases.
- The government has recognised 6817 oil spills in the region since the beginning of oil production. Critics say the number is much higher.
- Construction and increased ship traffic have changed local wave patterns, causing shore erosion and the migration of fish into deeper water.
- Various types of construction have taken place without adequate environmental impact studies.

The federal environmental protection agency has only existed since 1988 and **environmental impact assessments** were not compulsory until 1992. An environmental impact assessment is a document required by law detailing all the impacts on the environment of an energy or other project above a certain size.

Local people who have been forced to give up fishing because of reduced fish stocks often find it difficult to get alternative employment. Many local people feel that most jobs go to members of the country's majority ethnic groups – the Igbo, Yoruba, Hausa and Fulani who traditionally come from

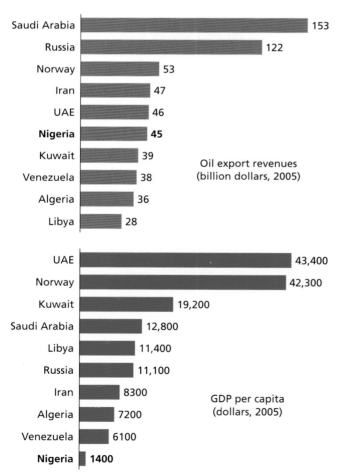

Figure 11 Nigeria's paradox

Figure 12 Environmental problems in the Niger delta

elsewhere in Nigeria. An added problem is the history of ethnic rivalry in an area inhabited by more than 20 ethnic groups. The people of the Niger delta also accuse the government of inadequate investment in the region in terms of schools, hospitals, housing and other forms of infrastructure. From time to time local rebel groups have attacked the oil industry either out of frustration at the paucity of benefits accruing to the region or in an attempt to gain payouts.

The largest new development in the delta is the Gbaran Integrated Oil and Gas Project operated by Shell. It is located along the Nun River, a tributary of the Niger. Encompassing 15 new oil and gas fields it should begin production in 2008. Shell hope to avoid the mistakes of the past with this project, which affects 90 villages.

Activities

1 Why are foreign companies so active in Nigeria's oil-fields?

2 Discuss the assertion that 'oil wealth has brought more disadvantages than advantages to the average person in Nigeria'.

▪ Tourism: risks and benefits

All tourism destinations are subject to both benefits and risks. However, the risks are greatest where the natural environment is extremely fragile. Two tourism destinations that fit into this category are the Galapagos Islands and the Great Barrier Reef.

Case Study

Ecuador's Galapagos Islands at risk

The Galapagos Islands, one of the most special environments in the world, are the most important tourism destination in Ecuador. The tourism industry brings in much-needed foreign currency and provides valuable employment opportunities in communities where alternative employment opportunities are limited. The government and the population in general clearly recognise the benefits of tourism, but recently major concerns have come to the surface.

In early 2007 the government of Ecuador declared the Galapagos Islands at risk, warning that visitor permits and flights to the island could be suspended. The Galapagos Islands straddle the equator about a thousand kilometres off the coast of Ecuador. All but 3 per cent of the islands are a National Park. Five of the 13 islands are inhabited. Visitor numbers are currently 100,000 a year and rising.

The volcanic islands can be visited all year round but the period between November and June is the most popular. A National Park entrance fee of £65 is payable on arrival. Among the many attractions are giant tortoises, marine iguanas and blue-footed boobies.

In signing the emergency decree to protect the islands, the President of Ecuador stated: 'We are pushing for a series of actions to overcome the huge institutional, environmental and social crises in the islands'.

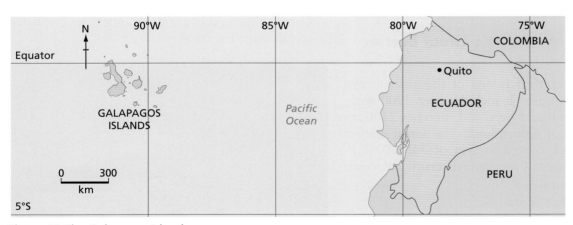

Figure 13 The Galapagos Islands

Figure 14 Tourists visiting the Galapagos Islands

The identified problems include:

- a growing population – 18,000 islanders with legal status earn a living from fishing and tourism but an additional 15,000 people are believed to live illegally in the islands
- illegal fishing of sharks and sea cucumbers which is believed to be at an all-time high
- the number of cruise ships which continues to rise
- internal arguments within the management structure of the National Park
- the controversial opening of a hotel in 2006.

Ecuador will monitor the environmental threat to the Galapagos Islands very closely in the future. The objective is to ensure that tourism is truly sustainable while maintaining clear economic benefits for those directly involved in the industry and for the country as a whole.

Case Study

The threat to Australia's Great Barrier Reef

The Great Barrier Reef is one of the great tourist attractions in Australia. It includes over 2900 reefs, around 940 islands and cays, and stretches 2300 km along the coast of Queensland. The Great Barrier Reef Marine Park covers an area of 345,000 km². This is an ecosystem of immense diversity including:

- 1500 species of fish
- 359 types of hard coral
- one-third of the world's soft corals
- six of the world's seven species of threatened marine turtle

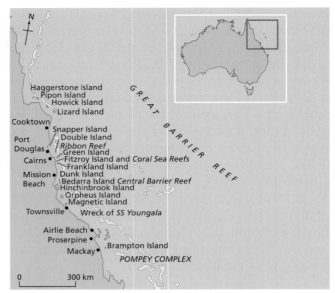

Figure 15 The Great Barrier Reef

- more than 30 species of marine mammals
- 215 bird species.

It has a significant economic impact on the state of Queensland (Figure 15) and in particular on the coastal resorts where tourists stay when visiting the reef. Economic activity based on the Great Barrier Reef contributed $5.8 billion to the Australian economy in 2004 and employed over 60,000 people. However, visitor pressure has been so great that the reef has suffered considerable damage.

There are several significant problems relating to the Barrier Reef:

- The impact of land-based pollution from agriculture, industry, residential areas and tourism is causing significant damage to the reef ecosystem.
- Overfishing – the use of dragnets in particular can damage the coral.

Figure 16 Coral on the Great Barrier Reef

- Coral bleaching is exacerbated by increased sea temperatures due to global warming. This causes coral polyps to die. As a result the range of colours is lost leaving only the white 'skeleton' of the coral.
- Tourists visiting the reef are causing damage by anchors, by reef walkers and divers and by pollution from the tourist boats visiting the area.

Until recently only 4.6 per cent of the reef was fully protected. After pressure from WWF and other organisations, the Australian government produced a plan to protect 33 per cent of the reef. The Great Barrier Marine Park zoning plan was implemented in 2004. It includes a network of marine sanctuaries, protecting over 11 million hectares, along the length of the reef.

A recent initiative has established new guidelines for tour companies using the reef. The objective is to make tourism more sustainable here. At present the scheme is voluntary but more and more companies are realising the benefits of acting in an environmentally friendly way.

Transport: risks and benefits

New, large-scale transport developments have become increasingly controversial as environmental groups and others highlight the direct and indirect problems associated with such developments. The largest-scale current transportation issue in the UK is the planned expansion of Heathrow airport. However, airport expansion is occurring or is planned in many other countries as well. The arguments for and against are broadly similar to those affecting Heathrow.

Case Study

The expansion of Heathrow airport

Heathrow, located 21 km west of central London, is the world's busiest international airport. More than 90 airlines fly to over 170 international destinations from this important hub airport. In 2007 almost 68 million passengers passed through the airport's terminals on 476,000 flights. Heathrow accounts for almost 30 per cent of all UK air passenger traffic. More than 35 per cent of Heathrow's passengers are business travellers. Heathrow is the second busiest airport in the world for cargo. It accounts for 55 per cent of all UK air freight by volume.

Recently a fifth terminal was opened at Heathrow which drew a number of protests, but the really big issue is the proposed third runway which would substantially expand the number of flights using Heathrow.

The main arguments against the further expansion of the airport are:

- the considerable increase in the number of people who will be affected by aircraft noise
- the increased noise levels for many people already affected by aircraft noise
- rising air pollution levels due to a considerable increase in flights
- a significant increase in road traffic generated by the extra flights
- the impact on wildlife.

Figure 17 shows the large area currently affected by aircraft noise from Heathrow. This map was published by the Department for the Environment, Food and Rural Affairs in late 2007. The area,

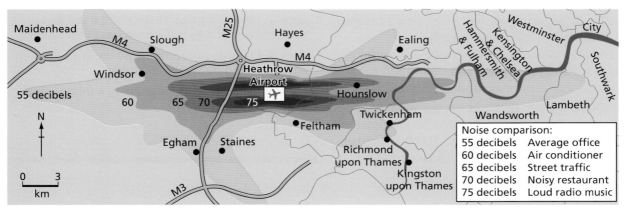

Figure 17 Heathrow airport with surrounding noise levels

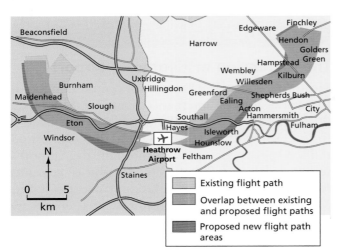

Figure 18 Existing and proposed new flightpaths at Heathrow

Figure 19 At Heathrow airport

stretching from the southern outskirts of Maidenhead in the west to the edge of Camberwell in the east, is home to 600,000 people who are affected by noise levels of 55 decibels or over. People living close to the airport are affected by noise levels of 75 decibels. Significant annoyance from aircraft noise begins at 50 decibels.

Campaigners against the expansion of Heathrow argue that plans to increase the number of flights from 420,000 a year to 700,000 will bring far more people within the area affected by aircraft noise. A recent study has highlighted the link between exposure to noise and ill health, noting in particular exposure to night-time aircraft noise and high blood pressure. The latter can lead to heart attacks and strokes.

The map in Figure 18, published in 2007, shows the changes to take-off and landing flightpaths that a third runway would bring. Many of the London boroughs affected argue that the government has not thought through the expansion plans carefully enough. The biggest impact will be in west London where all the flightpaths come together.

However, the economic importance of Heathrow to the local region, to London as a whole and to the national economy cannot be underestimated:

- Heathrow is a huge direct employer with 72,000 people working at the airport. It has been estimated that Heathrow supports another 100,000 further jobs in the UK. Nearly half of all those employed at

Heathrow live in the five boroughs directly surrounding the airport. Heathrow is the biggest single site employer in the UK.

- A large number of independent firms depend on Heathrow for much of their business. Examples are in-flight catering, security services, etc.
- Many companies say that relatively easy access to Heathrow was an important factor in locating in the surrounding region. Heathrow provides access to virtually every major city in the world. The airport has been described as 'the UK's gateway to the global economy'.
- There is a strong relationship between Heathrow and the financial services industry operating in the City of London.
- Supporters of the airport's expansion argue that if the third runway is not built, Heathrow will lose business to competing airports such as Paris and Amsterdam.

Activities

1 Look at Figure 17 (page 161). Describe the area currently affected by noise from Heathrow airport.

2 With the help of Figure 18, detail the arguments against a third runway at Heathrow.

3 Outline the arguments for building a third runway at Heathrow.

Geographical Skills and Investigations

Geographical skills

◼ Scale

Most topographical maps (sometimes called Ordnance Survey or OS maps) that we use are either at a 1:50,000 or a 1:25,000 scale. On a 1:50,000 map, 1 cm on the map relates to 50,000 cm on the ground. On a 1:25,000 map every 1 cm on the map refers to 25,000 cm on the ground. In every kilometre there are 100,000 cm (1,000 m × 100 cm). Hence:

- on a 1:50,000 map every 2 cm corresponds to 1 km
- on a 1:25,000 map every 4 cm corresponds to 1 km.

A 1:25,000 map is more detailed than a 1:50,000 map and is therefore an excellent source for geographical enquiries. 1:50,000 maps provide a more general overview of a larger area. You may come across other scales, e.g. 1:10,000 and 1:2,500.

Measurement on maps is made easier by grid lines. These are the regular horizontal and vertical lines you can see on a topographical map.

The horizontal lines are called **northings** and the vertical lines are called **eastings**. They help to pinpoint the exact location of features on a map.

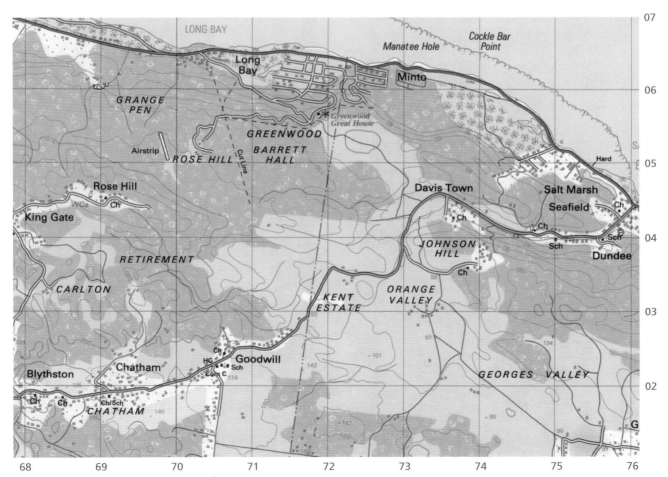

Figure 1 Part of the 1:50,000 map of Jamaica

Grid and square references

Grid references are the six-figure references which locate precise positions on a map. The first three figures are the eastings and these tell us how far a position is across the map. The last three figures are the northings and these tell us how far up the map a position is. An easy way to remember which way round the numbers go is 'along the corridor and up the stairs'.

In Figure 1 (page 163), the church at Rose Hill is located at 691046 and the Jetty at Dundee is found at 765044.

Sometimes a feature covers an area rather than a point, for example all of the villages and the areas of woodland in Figure 1. Here a grid reference is inappropriate so we use four-figure square references:

- The first two numbers refer to the eastings.
- The last two numbers refer to the northings.
- The point where the two grid lines meet is the bottom left-hand corner of the square.

Thus in Figure 1, most of the village of Seafield is found in 7504. Some features may occur in two or more squares, for example Long Bay is found in squares 7006 and 7106.

Direction

Directions can be expressed in two ways:

- by compass points, e.g. south-west
- by compass bearings or angular directions, e.g. 045°.

Sixteen compass points are commonly used. Some of these are shown in Figure 2.

Compass bearings are more accurate than compass points but can be quite confusing. Compass bearings show variations from magnetic north. This is slightly different from the grid north shown on a topographical map (which is the way in which the northings go). True north is different again – this is the direction of the North Pole.

Activities

Study Figure 1 (page 163) and look at the key to the Map on page 34.

1 How far is it:

   a) in a straight line
   b) by road

   from the school in Goodwill to the school in Dundee?

2 What is the length of the coastline (to the nearest km) as shown on the map extract?

3 Approximately how long is the airstrip?

4 How wide is the mangrove forest between Minto and Salt Marsh?

5 What is the six-figure grid reference for:

   a) the two schools at Dundee
   b) Greenwood Great House?

6 What is found at 719057?

7 Give the four-figure grid references for Chatham and Davis Town.

8 Suggest reasons why there is an airstrip in 6905.

9 In what direction is:

   a) Long Bay from Davis Town
   b) Goodwill from Rose Hill?

Relief and gradient

Contour lines

A **contour line** is a line that joins places of equal height.

- When the contour lines are spaced far apart the land is quite flat.
- When the contour lines are very close together the land is very steep (when the land is too steep for contour lines a symbol for a cliff is used).
- When contour lines are close together at the top, and then get further apart, it suggests a concave slope.
- When contour lines are close at the bottom and flat at the top, it suggests a convex slope.

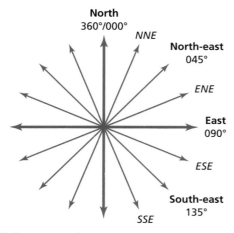

Figure 2 Compass points

Gradients

The gradient of a slope is its steepness. We can get a rough idea of the gradient by looking at the contour pattern. If the contour lines are close together the slope is steep, and if they are far apart the land is quite flat. However, these are not very accurate descriptions. To measure gradient accurately we need two measurements:

- the vertical difference between two points (this can be worked out using the contour lines or spot heights)
- the horizontal distance between two places – this may not be a straight line (for example, a meandering stream would not be straight).

Working out gradient

Make sure that you use the same units for both vertical and horizontal measurements.

Divide the difference in horizontal distance (D) by the height (H). If the answer is, for example, '10' express it as '1:10' ('one in ten'); or '5' as '1:5' ('one in five'). This means that for every 10 metres along you rise 1 metre, or for every 5 metres in length the land rises (or drops) 1 metre.

Alternatively, divide the height (H) by the difference in horizontal distance (D) and multiply by 100 per cent (H/D × 100%). This expresses the gradient as a percentage.

■ Describing river landscapes

The long profile of a river can be shown on a line graph when the height of a river above base level is plotted against distance from its source. As rivers evolve through time and over distance streams pass through a series of distinct changes. Figure 4 (page 167) shows the long profile of a river and illustrates these stages.

Describing the stages of the river

- Is the river in its upper, middle or lower course?
- Use the contour lines to describe the shape of the river valley – a V-shaped valley with close contour lines suggests the upper course; more gentle slopes with a broad, flat floodplain suggests the lower course.
- Look at the size and shape of the river channel.
- Is the channel constrained by relief, for example does it flow around interlocking spurs?
- Does the river meander across a flat floodplain?
- What are the features of the river? Can you identify any of the features listed in Figure 4 from map evidence?

Rivers have had a profound effect on both the site and situation of human settlements. Human activities have also had an increasing impact on drainage basins and river channels. Map evidence can be used to identify the relationship between rivers and human activities.

Activities

Study Figure 3 (page 166).

1 What is the height of:

   a) Silver Hill (8658)
   b) Baker Hill (8455)?

2 In what direction does Little Bay face?

3 How steep is the slope between Silver Hill and the coastline at Thatch Valley? Measure from the peak of Silver Hill to the nearest point of the coast in Thatch Valley. Express your answer as a 'one in x slope'.

   (Note: The contours on this map are drawn at 50 foot intervals. Assume that 3 feet equals 1 metre.)

4 Describe the relief (height and gradient) of squares 8658 (Silver Hill), 8457 (Potato Hill) and 8655 (Judy Piece).

5 Following an eruption of the Soufrière volcano in 1997, much of the southern third of the island was evacuated. Plans were made to develop the northern part of Montserrat. Study the map and comment on:

   a) the problems of developing the northern part of the island
   b) which area, in your opinion, is the best location to develop housing, services and economic activity. Give reasons for your answer.

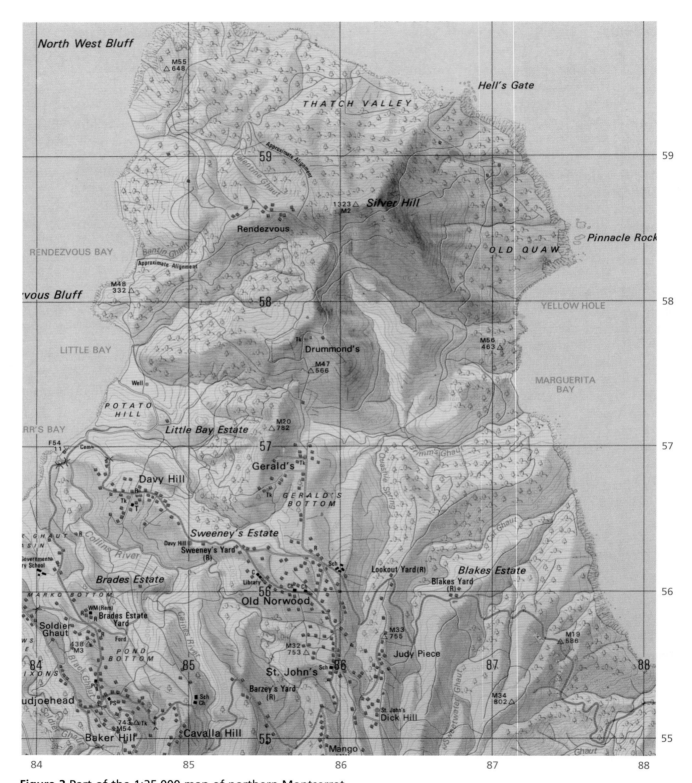

Figure 3 Part of the 1:25,000 map of northern Montserrat

Describing a river's influence on site and situation

- Is the river navigable? (Is it straight and wide enough to allow boats to pass up it?)
- Does the river valley provide the only flat land in an area of rough terrain?
- Does the valley provide a natural routeway for roads and railway lines?
- Do settlements avoid the river's floodplain and locate on higher dry-point sites?
- Are settlements located at crossing points on a river? (The name endings of some settlements, such as 'ford' and 'bridge', are evidence of this.)

The human impact on river systems

- Is there evidence of forest clearance and wetland reclamation for agriculture?
- Does the map show any of the following land-use changes which can affect a river and its drainage basin: mining activity, industrialisation, urbanisation, land drainage schemes?
- Has there been any direct interference with rivers through reservoir construction, channel straightening, dams, new channels?
- Are there any obvious sources of pollution (industry, sewage works) on the map?

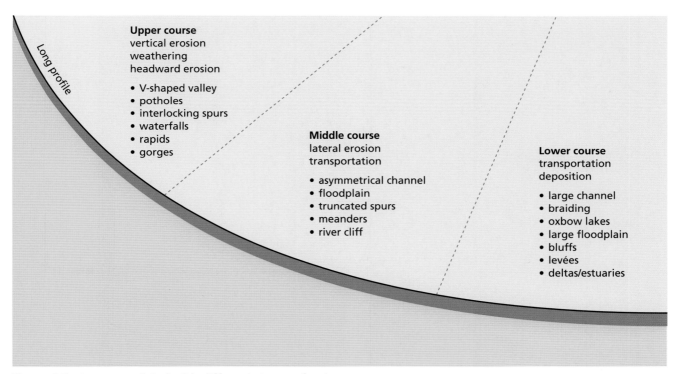

Long profile

Upper course
vertical erosion
weathering
headward erosion

- V-shaped valley
- potholes
- interlocking spurs
- waterfalls
- rapids
- gorges

Middle course
lateral erosion
transportation

- asymmetrical channel
- floodplain
- truncated spurs
- meanders
- river cliff

Lower course
transportation
deposition

- large channel
- braiding
- oxbow lakes
- large floodplain
- bluffs
- levées
- deltas/estuaries

Figure 4 Features associated with different stages of a river

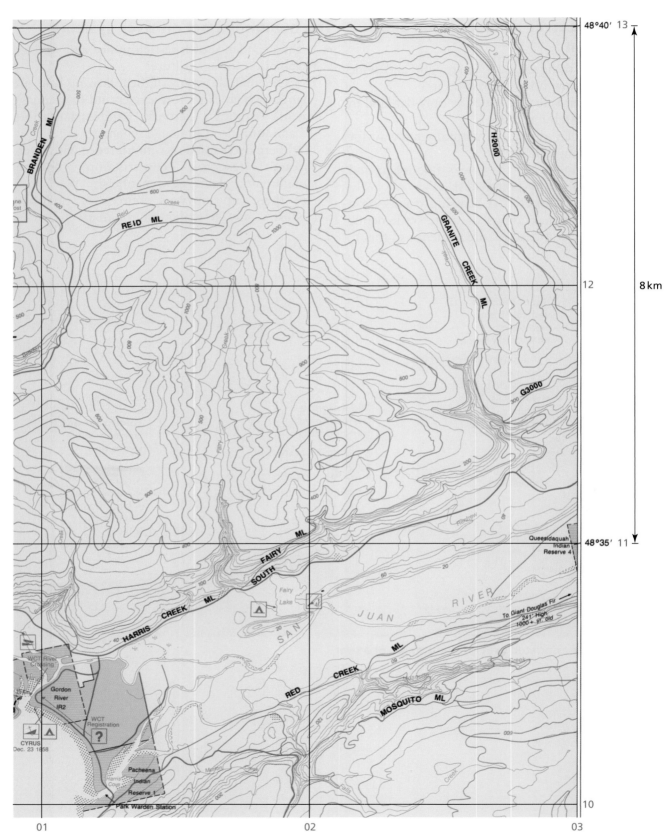

Figure 5 Part of the West Coast Trail in British Columbia, Canada

Activities

Study Figure 5.

1 Describe the relief (height and gradient) of the map extract.

2 In which direction does Fairy Creek flow?

3 a) Describe the valley of the San Juan River.
  b) How does this compare with the southern part of the Fairy Creek?

4 a) What is the altitude (height) of the source (start) of Fairy Creek?
  b) What is its altitude when it reaches the Fairy Lake?
  c) What is the distance from the source of Fairy Creek to Fairy Lake?
  d) What is the gradient of Fairy Creek between its source and the Fairy Lake? (Express the answer as a 1:x gradient, where gradient = vertical difference/horizontal distance.)

Coastal landforms

Describing coastal scenery

- Does the coastline have steep slopes and cliffs, suggesting a coastline of erosion? Or are there wide expanses of sand and mud suggesting deposition?
- Are there many headlands and bays indicating local changes in processes?
- Is the coastline broken by river mouths or estuaries?
- What is the direction of the coastline?
- Is there any evidence for longshore drift, e.g. spits, bars, tombolos?
- Are any of the features named? Give names and grid references.
- Is there any map evidence of human attempts to protect the coastline, e.g. groynes, seawalls, breakwaters?
- Does the map tell you whether the stretch of coastline is protected or open?

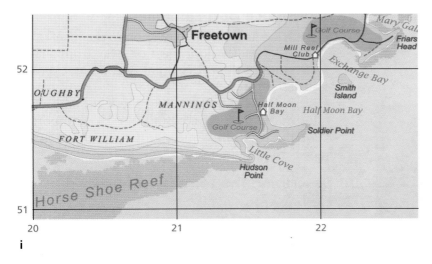

i

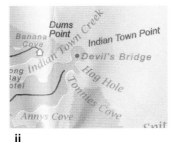

ii

iii

Figure 6 Extracts from the 1:35,000 map of Antigua

a

b

c

Figure 7 Some coastal landforms in Antigua

Rural settlement

Geographers should be able to find various type of information when studying rural settlements on a map. They should be able to comment on the site and situation of a village, its form (shape) and function, and the general distribution of settlements on a map.

> ### Key terms
>
> **Site:** the immediate location of a settlement – that is, the land on which it is built (e.g. on a floodplain, close to a river, on a south-facing slope, on a crossroads, wet point, dry point).
>
> **Situation:** the relative location of a settlement to a larger area.
>
> **Function:** any service or employment opportunities that a settlement offers (e.g. commercial, recreational, industrial, agricultural, etc.).
>
> **Shape:** the appearance of the settlement (e.g. linear, compact, T-shaped).

Describing and explaining the site and importance of a settlement

- Describe its location in relation to the relief of the area (e.g. valley floor, near or away from a river, direction of slope, dry point, wet point, exposed, sheltered, etc.)
- Describe how many routes there are and which forms of transport are available. How important are the routes that meet at the settlement? How does the relief effect these routes?

Describing the form or shape of a settlement

- Is it nucleated or dispersed?
- Is it a linear or cruciform settlement?
- Is any of it modern? (Modern settlement may be recognised by a regular geometric street pattern and more widely-spaced houses.)

Activities

1 Match each of the map extracts in Figure 6 (page 169) with one of the photographs in Figure 7.

2 Try to identify the cliff shown on Figure 7a. What is the map evidence to support your answer?

3 What are the features shown in Figure 7b? Find examples of these features on the map extracts and state their location.

4 What type of feature is shown in Figure 7c? Try to find a named example of the feature on one of the map extracts.

5 What is the map evidence to suggest that tourism and recreation are important to this area?

6 Using map evidence, suggest how easy or difficult it may be to develop tourism further in the area.

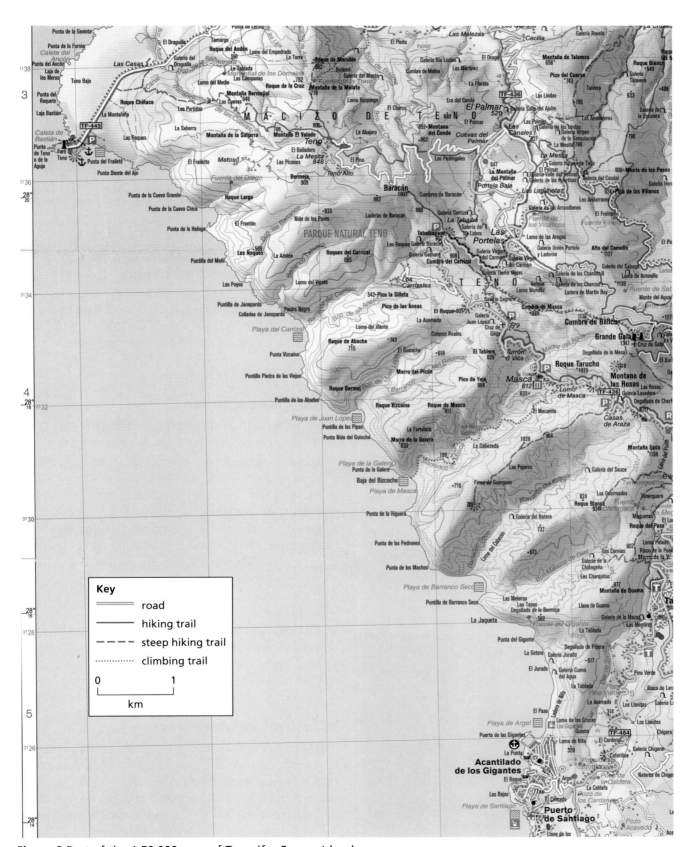

Figure 8 Part of the 1:50,000 map of Tenerife, Canary Islands

- Have physical features influenced its shape? (e.g. A steep hillside or floodplain may limit growth of a town so that it becomes elongated, following the direction of the level or dry land.)

Describing the size and function of a settlement

- What is the size of the settlement? How many grid squares does it cover? (Check what area each grid square represents, e.g. 1km².)
- Are the houses tightly packed or dispersed?
- What functions are evident? For example, is it residential (housing), commercial (post office, administrative buildings), schools, industrial (works, quarries, railway sidings), and/or tourist-related (tourist information centre, viewpoints)?

Describing the situation of a settlement

- Using the whole map area, describe the site of the settlement in relation to large urban centres, motorways, roads, large rivers and other large-scale physical features (such as hills or valleys).
- Is the settlement on a rail link?
- How accessible is the settlement to motorways, railways and large urban areas? It might be situated close to a motorway but have no direct access to it.

Describing and accounting for the general distribution of settlement

- Locate areas with little or no settlement. Account for the lack of settlement in terms of natural disadvantage (e.g. exposed position, steep gradient, flat land in danger of flooding). How is this land used?

Figure 9 Masca, Tenerife

- Locate areas of fairly close settlement. Account for this in terms of natural advantages for land use and occupation (e.g. farming, water, soil, south-facing slope).

Activities

Study Figure 8 (page 171).

1 Describe the pattern of settlement (grey squares) as shown on the map extract.

2 Describe the relief and gradient of the area west of Masca.

3 Describe the road network in the map extract.

4 What opportunities does this environment offer? Suggest reasons and use the map evidence and Figure 9 to help you.

5 What difficulties does this environment create? Again give evidence and suggest reasons.

Urban settlement

Urban landscapes

It is important to identify different land-use types when describing an urban landscape.

- Identify the lines of communication (e.g. roads, railways, airports).
- Identify the types of land use (e.g. residential, industrial, commercial). What kind of residential land use is it? Is it terraced, semi-detached, working class, middle class? Is there industry present? If so, what kind? Can you see a CBD or a grouping of shops?
- If the map is of a whole settlement, can you identify old and new areas?
- Use descriptive words when discussing the area: high-density/low-density housing; regular/haphazard roads; derelict/high-tech industry.
- Can you identify any other land uses (e.g. green spaces, derelict land, churches, schools, hospitals)?

Industrial location

The factors determining industrial location are changing. In the early part of the twentieth century, heavy industries like iron and steel and car manufacturing were located close to raw materials and/or markets. Today, manufacturing industry is drawn to out-of-town or edge-of-town sites. One of the factors is space and another is cost. There is a greater amount of cheaper land available away from the built-up areas of urban areas. Another factor is accessibility – edge-of-town sites are closer to communications (motorways and railways) and the residential areas that workers live in. For many heavy industries, location by a deepwater channel is important for the import of raw materials and the export of finished goods.

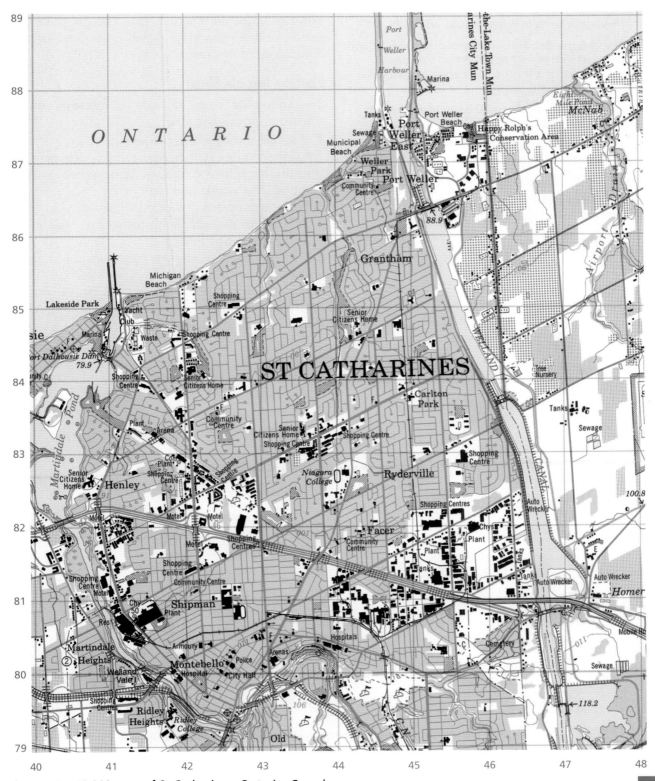

Figure 10 1:50,000 map of St Catharines, Ontario, Canada

Activities

Study Figure 10 (page 173).

1 Describe the site of St Catharines.

2 Suggest contrasting reasons for the lack of settlement in parts of squares 4081 and 4379.

3 a) Describe the distribution of industries (e.g. 'Plant' and 'Auto Wrecker') as shown on the map.
   b) Suggest reasons for the large-scale industry in squares 4581 and 4681.

4 Describe the distribution of shopping centres, as shown on the map.

5 Contrast the pattern of roads in grid square 4280 with those in 4083 and 4184.

6 Give four-figure grid references for:
   a) Martindale Pond
   b) the Port Weller Harbour.

7 Give a six-figure grid reference for Niagara College (near the centre of St Catharines).

■ Pie charts and bar charts

Pie charts

Pie charts are subdivided circles. These are frequently used on maps to show variations in composition of a geographic feature (Figure 11). The pie chart may also be drawn proportional in size, to show an extra dimension, in this case the size of GRDP.

Plotting the pie chart

The following steps should be followed in the construction of a pie chart.

1 Convert the data into percentages.
2 Convert the percentages into degrees (by multiplying by 3.6 and rounding up or down to the nearest whole number).
3 Draw appropriately located circles on your map.
4 Subdivide the circles into sectors using the figures obtained in step 2.
5 Differentiate the sectors by means of different shadings or colours.
6 Draw a key explaining the scheme of shading and/or colours.
7 Give the diagram a title.

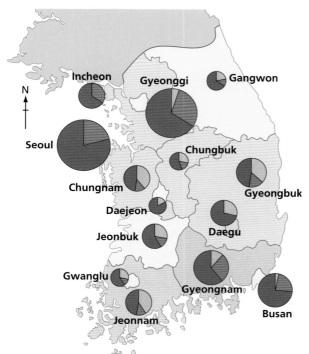

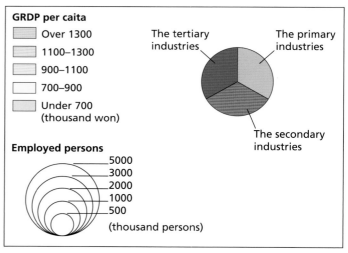

Figure 11 Example of the use of pie charts: employment and gross regional domestic product (GRDP) in South Korea, 2000

Bar charts

In a bar chart, the length of the bar represents the quantity of each component (e.g. places or time intervals). The vertical axis has a scale which measures the total for each of these components. There are four main types of bar chart, as follows.

- **Simple bar chart** – each bar indicates a single factor. If the difference in length of bars is not great, it can be emphasised by leaving a space between them or breaking the vertical scale.
- **Multiple or group bar chart** – features are grouped together on one graph to help comparison.
- **Compound bar chart** – various elements or factors are grouped together on one bar (the most stable element or factor is placed at the bottom of the bar to avoid disturbance).
- **Percentage compound bar chart** – this is a variation on the compound bar chart. It is used to compare features by showing the percentage contribution. These graphs do not give a total in each category but compare relative changes in percentages.

Activities

Year	Total	USA	Japan	UK
1985	532.2	108.0	364.3	12.3
1995	1947.2	644.9	418.3	86.7
2000	15,216.7	2922.0	2448.0	84.0
2005	11,563.5	2689.8	1878.8	2307.8

Figure 12 Foreign investment into Korea, US $ million

1 **a)** Using the data in Figure 12, draw a compound bar graph to show how foreign investment varied between 1985 and 2005. Draw one bar for each year.
   b) Describe the changes as shown in your bar chart.

Year	Automobile manufacturing (thousand vehicles)	Shipping orders (thousand gross tonnes)	Steel manufacturing (thousand gross tonnes)
2005	3699	19,279	56,306

Figure 13 The principal manufacturing products in Korea, 2005

2 **a)** Use the data in Figure 13 to draw a pie chart showing the principal manufacturing products in Korea in 2005.
   b) Describe the results you have produced.

■ Scatter graphs and triangular graphs

Scatter graphs

Scatter graphs show how two sets of data are related to each other, for example population size and number of services, or distance from the source of a river and average pebble size. To plot a scatter graph, decide which variable is independent (population size/distance from the source) and which is dependent (number of services/average pebble size). The independent variable is plotted on the horizontal or x axis and the dependent on the vertical or y axis. For each set of data project a line from the corresponding x and y axis (see Figure 14) and where the two lines meet a dot or an x is marked.

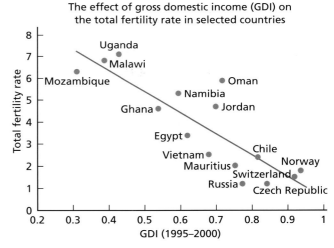

Figure 14 A scatter graph

Triangular graphs

Triangular graphs are used to show data that can be divided into three parts, e.g. soil (sand, silt and clay), employment (primary, secondary and tertiary) and population (young, adult and elderly) (Figure 15, page 176). Data for the graph should be in percentage figures and the percentages must total 100. The main advantage of a triangular graph is that it allows:

- a large number of data to be shown on one graph (think how many pie charts or bar charts would be needed to show the data on Figure 15)

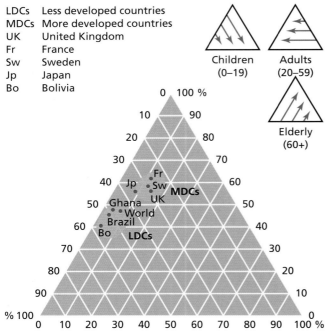

LDCs   Less developed countries
MDCs   More developed countries
UK   United Kingdom
Fr   France
Sw   Sweden
Jp   Japan
Bo   Bolivia

Figure 15 Triangular graph to show population composition in selected countries

- groupings to be easily recognisable – in the case of soils, groups of soil texture can be identified
- dominant characteristics to be shown easily
- classifications to be drawn up.

Triangular graphs can be tricky and it is easy to get confused when creating them. However, they provide a fast and reliable way of classifying large amounts of data that have three components.

■ Sketch maps and annotated photographs

You can label a photo or diagram to make it very informative. It is important that you label clearly all the relevant features.

Many photographs used in exams are aerial views which show industrial, residential, recreation and commercial land uses. In your projects, however, you are more likely to have much simpler photos. If you study these carefully you can find out a number of interesting features.

Activities

1 Construct a scatter graph using the following data:

Site	Discharge (m³/sec)	Suspended load (g/m³)
1	0.45	10.8
2	0.42	9.7
3	0.51	11.2
4	0.55	11.3
5	0.68	12.5
6	0.75	12.8
7	0.89	13.0
8	0.76	12.7
9	0.96	13.0
10	1.26	17.4

When all the data are plotted, draw a line of best fit. This does *not* have to pass through the origin. It is useful to label some of the points, for example the highest and smallest anomalies (exceptions) – especially if these are referred to in any later description.

2 On a copy of Figure 16, use the following data to show how the workforce of Korea has changed over time:

	Primary industries	Secondary industries	Tertiary industries
1970	50.4	14.3	35.3
1980	34.0	22.5	43.5
1990	17.9	27.6	54.5
2000	10.9	20.2	68.9

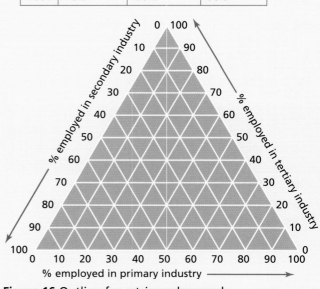

Figure 16 Outline for a triangular graph

Activities

Figure 17 Aerial view of Hyundai shipyard, Pusan, South Korea

1 Study Figure 17. Copy the sketch below the photo and add the following labels in the most appropriate locations.

Harbour wall to reduce wave energy
Flat land for large-scale industrial development
Lack of development on steep ground
High-rise residential accommodation
Large docks for ships to be repaired or built
Add a couple of other labels based on your own observations.

2 Study Figure 18. Make a sketch diagram of the resort and add the following labels:

Purpose-built holiday resort
Easy access to the beach
Boat moorings
Fine, white sandy beach
Bay
Lagoon

Figure 18 The Jolly Harbour resort in Antigua

Geographical investigations

Sequence of investigation

The enquiry skills for Paper 4 can be set out in the following stages:

- formulating aims and hypotheses
- data collection
- data presentation techniques
- data analysis
- formulation of conclusions
- evaluation.

Formulating aims and hypotheses

This section forms your introduction to the investigation. Here you should state the broad purpose of the study, its aims and location. Many geographical investigations begin by stating one or a number of hypotheses. Hypotheses are the ideas you intend to test. Before you can set out your hypotheses with confidence you need to ensure that you have a good understanding of the topic (for example, sand dunes) under consideration. Studying the 'geographical background' should ensure that you have a clear knowledge and understanding of the theories or models that are used to try to explain your enquiry. You will refer back to these theories and models in your conclusion.

Examples of hypotheses are:

- Pedestrian density is highest at the centre of the CBD, and declines with increasing distance from the centre.
- The sphere of influence of settlements increases with settlement size.
- The pH of sand dunes decreases with distance inland.
- Population density is higher in inner urban areas than in the suburbs.
- Average temperatures in urban areas are higher than in surrounding rural areas.

For each hypothesis you investigate you should describe what you expect to find and explain why. Within this section of the investigation you should also justify the geographical location of your inquiry. Make sure that you include the area's site as well as its regional situation. Include clearly labelled location maps. Give each map or diagram used in this section a figure number. For example, a map showing the location of your study area within its wider region might be labelled 'Figure 1: the location of Studland in Dorset'. Follow this procedure for all illustrations used throughout your investigation.

It is also useful in this opening section of your investigation to briefly state the sequence of investigation you are going to follow. This should ensure that you are clear about the remaining stages of the investigation.

Data collection

Use as many different techniques as possible to gather information, for example, interviews, observations, surveys, questionnaires, maps and looking at figures. Describe and justify each method. Some of these are primary (your own fieldwork) and some are secondary (published statistics, maps, diagrams, etc.). You need both. Describe the use of primary fieldwork methods and in particular the method or equipment used to collect each type of information. Equally, describe and explain the use of secondary sources, for example, the census, parish records.

Explain clearly how you decided to use your figures, maps, answers to questions, etc. Some reasoning is necessary here – that is, justify why you used that method or source. Explain in detail, for example, how you questioned people, collected census figures, obtained maps, etc. Write this up almost like the method for a scientific experiment. You can use a planning sheet here stating when you collected data, where from, at what time, places you visited, observations you made, interviews you conducted, etc. If you are using a questionnaire then you must justify the questions that you use, for example, why have you recorded the age and gender of respondents in a shopping survey? You need a range of methods to obtain full marks.

To collect data in a sound and logical way so that valid conclusions can be drawn you should be aware of the characteristics and importance of:

- sampling
- pilot surveys
- questionnaires and interviews
- methods of observing, counting and measuring
- health and safety and other restrictions.

Sampling and piloting

Reasons for sampling

For many geographical investigations it is impossible to obtain 'complete' information. This is usually because it would just take too long in terms of both time and cost. For example, if you wanted to study the shopping habits of all 1000 households in a suburban area by using a doorstep questionnaire, it would be a huge task to visit every household. However, it is valid to take a sample or proportion of this total 'population' of 1000 households, providing you follow certain rules. The idea is that you are selecting a group that is representative of the total population.

You might possibly decide to take a 5 per cent or 10 per cent sample which would involve talking to 50 or 100 of the 1000 households in the area. But how do you decide which 50 or 100 households to sample? There are three recognised methods of sampling which are considered scientifically valid. All three methods avoid bias which would make results unreliable.

Sampling types

Before selecting the sampling method you need to consider how you are going to take a sample at each location. The alternatives are:

- Point sampling – making an observation or measurement at an exact location such as an individual house or a precise six-figure grid reference.
- Line sampling – taking measurements along a carefully chosen line or lines such as a transect across a sand dune ecosystem.
- Quadrat (or area) sampling – quadrats are mainly used for surveying vegetation and beach deposits. A quadrat is a gridded frame.

All three sampling types are shown in Figure 1. Here all of the sampling types are illustrated using the systematic method of sampling. When you have read the next section you might think how these diagrams would look using random and stratified sampling.

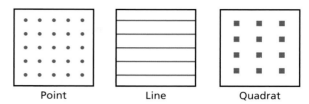

Figure 1 Point, line and quadrat sampling

Sampling methods

Random sampling

This method involves selecting sample points by using random numbers. Tables of random numbers can be used or the numbers can be generated by most calculators. The use of random numbers guarantees that there is no human bias in the selection process.

Systematic sampling

With this method the sample is taken in a regular way. It might, for example, involve every tenth house or person. When using a topographical map it might mean analysing grid squares at regular intervals.

Stratified sampling

Here the area under study divides into different natural areas. For example, rock type A may make up 60 per cent of an area and rock type B the remaining 40 per cent. If you were taking soil

61	89	04	24	98	65	96	96
33	79	53	35	51	56	11	78
96	84	68	33	84	15	08	10
28	34	05	81	54	02	60	18
19	35	37	56	39	97	66	15
37	21	22	09	18	99	33	03
46	77	77	83	19	39	43	48
12	44	97	58	79	57	42	30
08	91	47	87	38	21	74	24
98	17	54	62	62	21	06	90
73	53	29	99	11	76	30	00
35	28	06	62	12	99	48	48
50	34	68	74	61	42	19	63
95	49	75	96	49	81	93	10
22	30	86	92	56	79	71	50
68	83	63	59	30	55	37	20
69	67	64	05	14	37	16	36
04	43	66	24	01	62	72	98
03	40	89	99	66	22	11	32
95	44	09	92	08	41	49	27

Figure 2 Section of a table of random numbers

samples for each type, you should ensure that 60 per cent of the samples were taken on rock type A and 40 per cent on rock type B.

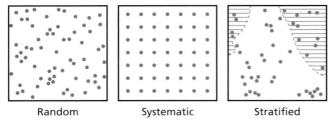

Random · Systematic · Stratified

Figure 3 Random, systematic and stratified sampling

Deciding on the size of the sample

The larger the sample the more likely you will obtain a true reflection of the total population. However, this could happen with a very small sample by chance. Equally a small sample could give a very misleading picture of the total population. A good rule to follow at IGCSE with regard to sample size is to take as many samples as possible with regard to:

- the time available
- available resources
- the number of samples required for a particular statistical technique such as Spearman's rank correlation coefficient
- your capacity to handle the data collected (there are many computer programs available to help with this).

Piloting

A pilot study or trial run can play an important role in any geographical investigation. A pilot run involves spending a small amount of time testing your methods of data collection. For example:

- If you are using equipment, does all the equipment work and can everyone in the group use it correctly?
- If your data collection involves a questionnaire, can the people responding understand all the questions clearly?
- If a method of sampling is used, does everyone know how to select the sample points accurately?

A small-scale pilot study allows you to make vital adjustments to your investigation before you begin the main survey. This can save a great deal of time in the long run.

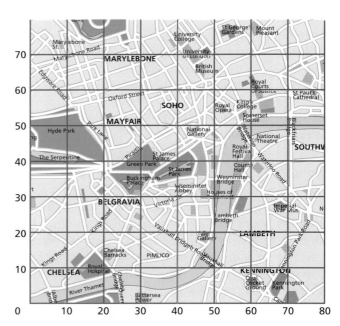

Figure 4 How a grid and map can be used for random sampling

Activities

1 Work in groups to provide outlines of different geographical investigations that would involve:
   a) random sampling
   b) systematic sampling
   c) stratified sampling.

2 Why might it be beneficial to conduct a pilot study prior to beginning a geographical investigation?

Questionnaires and interviews

Questionnaire surveys involve both setting questions and obtaining answers. The questions are pre-planned and set out on a specially prepared form. This method of data collection is used to obtain opinions, ideas and information from people in general or from different groups of people. The questionnaire survey is probably the most widely used method to obtain primary data in human geography. In the wider world questionnaires are used for a variety of purposes including market research by manufacturing and retail companies and to test public opinion prior to political elections.

One of the most important decisions you are going to have to make is how many questionnaires you are going to complete. The general rules to follow here are similar to those for sampling, set out in the previous section. Remember, if you have too few questionnaire results, you will not be able to draw valid conclusions. For most types of studies, 25 questionnaires is probably the minimum you would need to draw reasonable conclusions. On the other hand it is unlikely you would have time for more than 100 unless you were collecting data as part of a group.

A good questionnaire:

- has a limited number of questions that take no more than a few minutes to answer
- is clearly set out so that the questioner can move quickly from one question to the next (people do not like to be kept waiting – the careful use of tick boxes can help this objective)
- is carefully worded so that the respondents are clear about the meaning of each question
- follows a logical sequence so that respondents can see 'where the questionnaire is going' (if a questionnaire is too complicated and long-winded people may decide to stop halfway through)
- avoids questions that are too personal
- begins with the quickest questions to answer and leaves the longer/more difficult questions to the end
- reminds the questioner to thank respondents for their co-operation.

The disadvantages of questionnaires are:

- Many people will not want to co-operate for a variety of reasons. Some people will simply be too busy, others may be uneasy about talking to strangers, while some people may be concerned about the possibility of identity theft.
- Research has indicated that people do not always provide accurate answers in surveys. Some people are tempted to give the answer that they think the questioner wants to hear, or the answer they think shows them in the best light.

As with other forms of data collection, it is advisable to carry out a brief pilot survey first. It could be that some words or questions you find easy to understand cause problems for some people. Amending the questionnaire in the light of the pilot survey before you begin the survey proper will make things go much more smoothly.

Delivering the questionnaire

There are really three options here:

- Approach people in the street or in another public environment.
- Knock on people's doors.

- Post questionnaires to people. With this approach you could either collect the questionnaire later or enclose a stamped addressed envelope. The latter method is costly and experience shows that response rates are rarely above 30 per cent. Another disadvantage is that you will be unable to ask for clarification if some responses are unclear.

If you are conducting a survey of shopping habits you may want to find out if there are significant differences between males and females and between different age groups. In this case you would use a stratified sample divided by gender and the percentage of population in each age group.

The time of day may also be important. In the example given above, very few people in some age groups may be around at a certain time of day. For example, most teenagers will be in school or college at mid-morning on a weekday.

Interviews

Interviews are more detailed interactions than questionnaires. They generally involve talking to a relatively small number of people. A study of an industrial estate might involve interviews if you were trying to find out why companies chose to locate on the estate. An interview is much more of a discussion than a questionnaire although you should still have a pre-planned question sheet. It can be a good idea to record interviews but you should ask the interviewee's permission first.

Health and safety and other restrictions

It may be sensible to work in pairs when conducting questionnaires as some people can act in an unfriendly manner when approached in the street. Working in pairs can also speed the process up considerably, with one person asking the questions and the other noting the answers. Also be aware that shopping malls, individual shops and other private premises may not allow you to conduct questionnaires without seeking permission beforehand.

Activities

1 Design a questionnaire that might be used as part of an investigation into tourism in a small resort.

2 Briefly outline a geographical investigation in your local area that could involve the use of interviews.

A good questionnaire

Introduction: 'Excuse me, I am doing a school geography project. Could I ask you one or two quick questions about where you go shopping?'

1 How often do you come shopping in this town centre?

More than once a week ☐

Weekly ☐   Occasionally ☐

2 How do you travel here?

Walk ☐   Car ☐   Bus ☐   Train/Tube ☐
Other _____

3 Roughly where do you live? _____

4 Why do you come here rather than any other shopping centre?

Near to home ☐   Near to work ☐

More choice ☐   Pleasant environment ☐
Other _____

5 What sort of things do you normally buy here?

Groceries ☐   Clothes/shoes ☐

Everything ☐

Other _____

6 Do you shop anywhere else, and if so where?

_____

7 Why do you go shopping there?

_____

8 What do you buy there?

_____

9 Sex: M ☐ F ☐   Age (estimate) under 20 ☐
                        20–30 ☐ 30–60 ☐ Over 60 ☐

'Thank you very much for you help'

A bad questionnaire

Introduction: 'Excuse me, but I wonder if I could ask you some questions?'

1 Where do you live?

_____

2 How do you get here?

_____

3 Do you come shopping here often?

_____

4 Why do you come here?

_____

5 Do you buy high- or low-order goods here?

_____

6 Is this a good shopping centre and if so, why?

_____

7 Where else do you go shopping?

_____

8 Do you shop there because it is cheaper or nearer to your home?

_____

9 How old are you? _____

'Right, that's it then.'

Figure 5 Two questionnaires to find out people's shopping habits, conducted in a town centre: one good and one bad

Observations, counts and measurements

Field sketches

Personal observations or perceptions may form an important element of a coursework investigation. A field sketch is a hand-drawn summary of an environment you are looking at. In both urban and rural environments field sketching is a very useful way of recording the most important aspects of a landscape and noting the relationships between elements of such landscapes. The action of stopping for a period of time to sketch the landscape in front of you often reveals details that may not have been apparent from a quicker look.

Figure 6 is an example of a good field sketch. This sketch highlights the important geographical features of the landscape. Key features should be clearly labelled but make sure that your sketch map is not too cluttered. This will detract from the really important details. A good field sketch will be viewed as a higher-level technique by your coursework moderator.

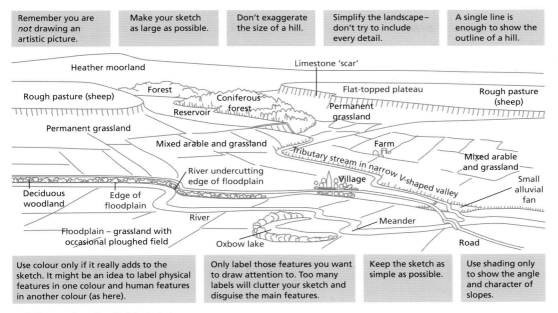

Remember you are *not* drawing an artistic picture.

Make your sketch as large as possible.

Don't exaggerate the size of a hill.

Simplify the landscape– don't try to include every detail.

A single line is enough to show the outline of a hill.

Heather moorland

Limestone 'scar'

Forest

Flat-topped plateau

Rough pasture (sheep)

Coniferous forest

Rough pasture (sheep)

Reservoir

Permanent grassland

Permanent grassland

Mixed arable and grassland

Farm

Tributary stream in narrow V-shaped valley

Mixed arable and grassland

River undercutting edge of floodplain

Village

Small alluvial fan

Deciduous woodland

Edge of floodplain

River

Meander

Floodplain – grassland with occasional ploughed field

Oxbow lake

Road

Use colour only if it really adds to the sketch. It might be an idea to label physical features in one colour and human features in another colour (as here).

Only label those features you want to draw attention to. Too many labels will clutter your sketch and disguise the main features.

Keep the sketch as simple as possible.

Use shading only to show the angle and character of slopes.

Figure 6 Example of a field sketch

Annotated photographs

Annotated photographs should be seen as complementing field sketches rather than being an alternative to them. Like field sketches, good, fully annotated photographs are regarded as a higher-level skill. Always record the precise location and conditions of the photographs you take. This should include grid reference, the direction the photograph was taken in, weather conditions and time of day. Such information will make annotation quicker and easier in the long run.

An annotated photograph shows your key perceptions about a location you have visited during fieldwork. A series of such photographs might show:

- how the type and quality of housing varies in an inner city or suburban area
- how a river and its valley change from source to mouth.

Annotations should be in the form of short, sharp sentences (Figure 7). Moderate abbreviation is fine providing the meaning of the comment remains

Flat, marshy flood plain.

Fields inundated as the river has topped its banks.

Road built on an embankment.

Limited number of trees to intercept rainfall.

Settlement built on slightly higher ground.

Figure 7 Example of an annotated photograph

clear. Some annotations will just be descriptive, but where the opportunity arises some explanation should also be included. Annotation can be most effective when the photograph is placed on the page in landscape format which will allow more space for annotations on all four sides. As with field sketches, a series of annotated photographs could form a very effective part of your analysis. You should look to correlate annotated photographs with the tables and graphs showing your data analysis. Photographs are also useful to show how you carried out surveys and field measurements.

Activities

1 a) Draw a field sketch of an urban or rural environment within easy reach of your school.
   b) Suggest why this location has geographical interest.

2 Annotate a photograph of a location of interest you have visited.

Recording tables

The most straightforward method of observation is noting whether a physical or human feature exists in an area or not. Figure 8 is an example of a recording table. The objective here is to compare the facilities in four parks before attempting to explain the differences between them. Recording is done by placing a tick in the appropriate square. Notice that there is a final column to accommodate any unexpected findings.

Scoring systems

Scoring systems are used in quality of life and other types of survey. Figure 9 is an example of a scoring system used to study variations in environmental quality in different parts of a residential area. Figure 9a shows that in this example ten local environmental factors are being observed. Figure 9b shows how the scoring system works. Here a score of 5 is the maximum possible for the best environmental conditions. The minimum score is 1. For each location the individual environmental scores are added together to achieve a total environmental score. In this example, the lowest possible total score is 10, and the highest is 50. It can be useful to practise the system in class using photographs before going out to conduct fieldwork.

Tally charts

Counts of various kinds are an element of many geographical investigations. Figure 10 is an example of a tally chart used to record visitor numbers at key locations in a park. The convention is to show counts in groups of five with the fifth count as a line drawn across the previous four counts.

You will notice that Figure 10 does not give a time when the count took place or state how long counting went on. In this example the number of visitors could vary significantly according to the time of day. It is therefore very important to plan carefully for your counts so that when you have collected and presented your data you can justify the conclusions you have drawn.

Park	Size (ha)	Woodland	Children's playground	Sports pitches	Bowling green	Tennis/basket-ball courts	Picnic site	Restaurant/café	Boating/fishing lake	Ornamental gardens	Pavillion/bandstand	Toilets	Car park	Info centre/gift shop	Other
High Lodge Forest Park	120	✓	✓				✓	✓				✓	✓	✓	Maze, jungle gym
Ditchingham Estate Park	25	✓							✓				✓		
Long Stratton Park	4	✓	✓	✓	✓	✓					✓	✓	✓		Skate ramp
Castle Mall Gardens	2						✓	✓		✓					Viewpoint

Figure 8 Example of a recording table showing park facilities

a

Ward name:		Location						
Factor	**1**	**2**	**3**	**4**	**5**	**6**	**7**	
1 Condition of brickwork/paintwork								
2 Condition of pipes, guttering, windows								
3 Quality/state of repairs of pavements								
4 Quality/state of repair of roads								
5 Extent of litter								
6 Extent of graffiti								
7 Presence/condition of vegetation								
8 Availability of parking								
9 General condition of front of house								
10 Age/number of vehicles								
Total:								
Ward total:								
Average Location Score:								

b

Explanation of the ranking system used for the environmental checklist		
Factor	**Explanations**	
1 Condition of brickwork/paintwork	**1** Worst condition	**5** Best condition
2 Condition of pipes, guttering, windows	**1** Worst condition	**5** Best condition
3 Quality/state of repairs of pavements	**1** Worst condition	**5** Best condition
4 Quality/state of repair of roads	**1** Worst condition	**5** Best condition
5 Extent of litter	**1** Most litter	**5** No litter
6 Extent of graffiti	**1** Large amount present	**5** None present
7 Presence/condition of vegetation	**1** Worst condition/none	**5** Best condition
8 Availability of parking	**1** None noticeable	**5** Space for 1+ car
9 General condition of front of house	**1** Worst condition	**5** Best condition
10 Age/number of vehicles	**1** Outdated/no vehicle	**5** New models/+1 vehicle

Figure 9 Example of an environmental scoring system

Pedestrian counts often form part of urban geography investigations. You could see how pedestrian counts decline with distance from the centre of the CBD. Pedestrian counts could be conducted every 50 or 100 m from this point.

	Children's playground	Sports ground	Bowling green
Male	ꟼꟼꟼ II	ꟼꟼꟼ ꟼꟼꟼ III	II
Female	ꟼꟼꟼ III		IIII

Figure 10 Example of a tally chart recording visitor numbers

Activities

1 Produce a recording table that could be used as part of a geographical investigation in your local area.

2 Look at the scoring system shown in Figure 9. Discuss the merits and limitations of this example.

■ Data presentation techniques

A wide variety of graphical techniques can be used to present geographical data. The skill is in choosing the best type of graph for the particular data set under consideration. Coursework marks can be lost by the incorrect use of graphical techniques. You should also consider the size of any graph or diagram you use. It is important that the labels of axes and all other information are clear and can be read easily.

It is important to integrate all maps, graphs, photographs and diagrams with the text. The most elementary way of doing this is to use a sentence such as: 'Figure 4 is a line graph showing temperature change in my garden.'

Line graphs

A line graph shows points plotted on a graph with the points connected to form a line. This type of graph is used to show continuing data. It shows the relationship between two variables which are clearly labelled on both axes of the graph. Many line graphs show changes over time. However, time does not have to be one of the variables of a line graph. Examples of the use of line graphs include:

- temperature changes during the course of a day
- pedestrian counts by time of day
- temperature change with altitude.

The axes of a line graph should begin at zero and the variable for each axis should be clearly labelled. Be careful with the choice of scale as this will determine the visual impression given by the graph. Figure 11 is an example of a line graph. Here, only one line has been drawn but it is valid to show a number of lines so long as the course of each line is absolutely clear from start to finish.

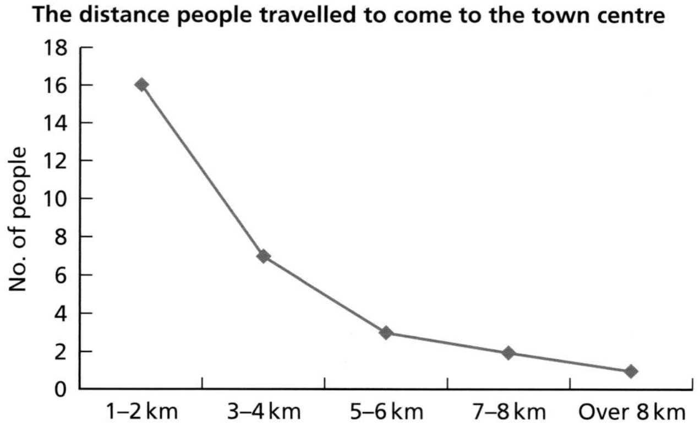

The distance people travelled to come to the town centre

Figure 11 Example of a line graph

Median-line bar graphs

Median-line bar graphs are useful when the objective is to show both positive and negative changes. The median line is set at zero, with the positive scale above the median line and the negative scale below it (Figure 12). This type of bar graph can create a very good visual impression. You can see instantly whether changes are positive or negative and exactly what the extent of the individual changes are.

Histograms

A histogram is a special type of bar graph (Figure 13). It shows the frequency distribution of data. The x axis must be a continuous scale with the values marked on it representing the lower and upper limits of the classes within which the data have been grouped. The y axis shows the frequency within which values fall into each of the classes.

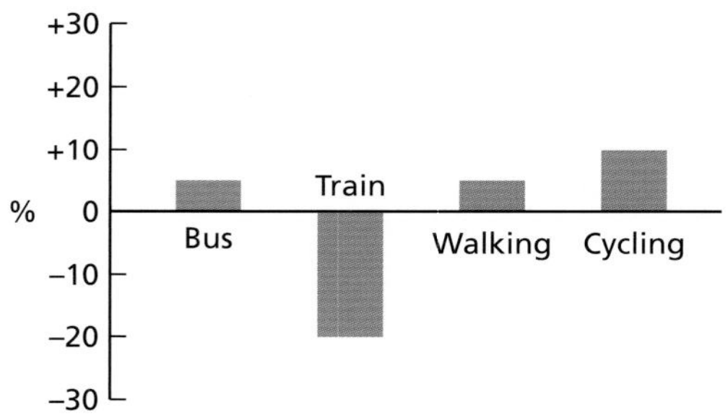

Percentage change in mode of travel to school, 2000–07

Figure 12 Example of a median-line bar graph

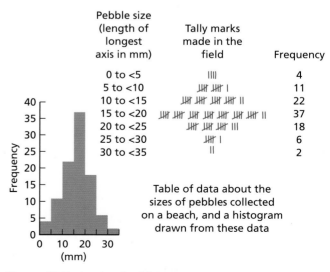

Pebble size (length of longest axis in mm)	Tally marks made in the field	Frequency
0 to <5	IIII	4
5 to <10	LHT LHT I	11
10 to <15	LHT LHT LHT LHT II	22
15 to <20	LHT LHT LHT LHT LHT LHT LHT II	37
20 to <25	LHT LHT LHT III	18
25 to <30	LHT I	6
30 to <35	II	2

Table of data about the sizes of pebbles collected on a beach, and a histogram drawn from these data

Figure 13 Example of a histogram

A vertical rectangle or bar represents each class. The bars must be continuous without any gaps between them.

Choropleth maps

Choropleth maps use variations in colour or different densities of black and white shading. The following steps should be followed in the construction of a choropleth map (Figure 14):

1 Look at the range of data and divide it into classes. There should be no less than four classes and no more than eight.
2 Allocate a colour to each class. The convention is that shading gets darker as values increase.
3 Now apply each colour to the relevant areas of the map.
4 Provide a key, scale and north point.

Proportional circles

Proportional circles are the next step up from pie charts. While pie charts are viewed as a basic graphical technique, proportional circles are a higher-level technique. Proportional circles are useful when illustrating the differences between two or more amounts. They are particularly effective when placed on location maps. In Figure 14 the three circles shown are proportional in area to the total number of offences recorded in the three urban areas. The method used to decide the radius of each circle is as follows:

1 In the first column write out each of the total figures for which circles are to be drawn.
2 Find the square root for each figure and write it down in the second column.
3 Use the square root for the radius of each circle (Figure 15). By doing this the area of each of the circles will be mathematically proportional to the figures they are representing. For the radii you can use any units you want providing they are the same for each of the circles.

Flow diagrams

Flow-line diagrams and maps are used to illustrate movements or flows. One might be used to show the different volumes of traffic from different smaller settlements into a larger settlement. Straight lines are used but the width of the individual flow lines will be proportional to the amounts of traffic they are representing. Thus a line 10 mm wide may represent 500 vehicles an hour along a road. On the same scale a line 2 mm wide would represent 100 vehicles an hour. Flow lines could also be used to show the number of buses coming into a town for a particular day.

Ray diagrams

There are two main types of ray diagrams: wind roses and desire lines. Ray diagrams comprise straight lines (rays) which show a connection or movement between two places.

Wind rose diagrams (Figure 16, page 188) show the variations in wind direction for a certain time period. The direction of each ray to the centre is the direction from which the wind is blowing. Each ray is proportional in length to the number of days the wind blew from that direction.

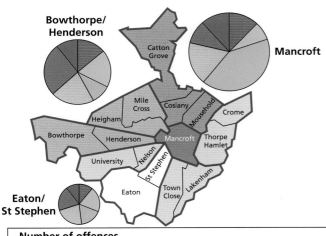

Number of offences committed
☐	Under 750
▢	750–999
▢	1000–1249
▨	1250–1499
▨	1500+
—	Boundary of beat areas

Types of crime
▨	Violent/sexual assult
▢	Burglary
▢	Theft
▢	Theft of/from motor vehicle (car crime)
▨	Criminal damage
▨	Other (including robbery)

Figure 14 Example of a choropleth map with proportional circles

Totals	Square root	Radius of circle
4	2	2 cm
9	3	3 cm
16	4	4 cm

Figure 15 Calculating radii for proportional circles

Wind directions recorded for one year at a school weather station in Liverpool, UK

Direction of wind	N	NE	E	SE	S	SW	W	NW	Calm
Number of days per year	26	37	39	32	30	57	60	53	51

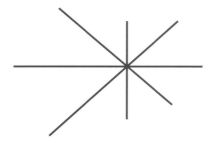

Figure 16 Example of a wind rose diagram

Desire line diagrams show movement from one place to another. This type of diagram could be used to show where people live and the supermarket they use. If there are four supermarkets in an area then the rays would focus on four points rather than just one as in a wind rose diagram. Desire line diagrams are therefore more complicated than wind rose diagrams.

Semantic differential profiles

Semantic differential profiles (SDPs) (Figure 17) are useful for recording perceptions of environmental quality. They consist of a series of pairs of words with opposite meanings. There should generally be a minimum of five gradations between each pair of words. The observer must decide where to place a cross or other mark to state the condition of the environment they are observing. When all the observations have been made the crosses are joined up with a ruler. If, for example you were studying three different housing areas in a town you could show all three profiles on one SDP by using different colours for each area.

Radial (circular) graphs

Radial (or circular) graphs (Figure 18) can be used to plot:
- a variable that is continuous over time, such as temperature data over the course of a year
- data relating to direction using the points of the compass.

The two axes of a radial graph are the circumference of the circle and the radius. Values increase from the centre of the circle outwards.

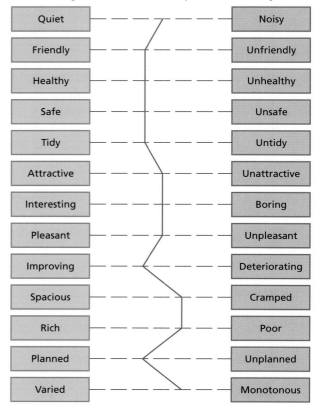

The average semantic differential profile for Graveny ward

Figure 17 A semantic differential profile

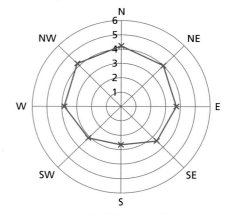

Rate of weathering: Rahn's index

A radial graph to show the influence of aspect on gravestone weathering. The graph shows the mean of Rahn's index for each compass direction.

Rahn's index

Class	Description
1	Unweathered
2	Slightly weathered; faint rounding to corners of letters
3	Moderately weathered: gravestone rough, letters legible
4	Badly weathered: letters difficult to read
5	Very badly weathered: letters indistinguishable
6	Extremely weathered: no letters left, scaling

Figure 18 A radial graph

Isoline diagrams

Isolines join points of equal value on a map. They are similar to contours on a topographical map. Isolines can only be drawn when the values under consideration change in a fairly gradual way over the area of the map. Data for quite a large number of locations is required in order to draw a good isoline map (Figure 19). Isoline maps are unsuitable for patchy data.

Activities

1 Find two examples of line graphs in a geography textbook. What do the line graphs show in these examples?

2 What is the difference between a histogram and an 'ordinary' bar graph?

3 Discuss the merits and limitations of choropleth maps.

4 Draw a series of proportional circles using fieldwork data or data from a textbook.

5 Construct a semantic differential profile that could be used in your local area.

6 Look at Figure 19. Suggest two more examples for which isoline maps could be drawn.

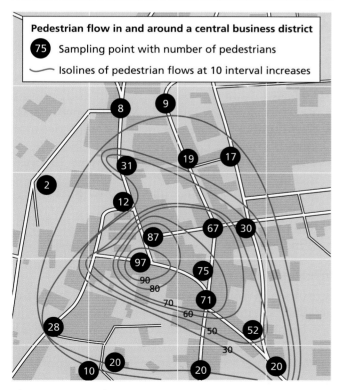

Figure 19 An isoline map

■ Data analysis

Here you should describe the patterns in data presented in graphs and tables of results:

● After each graph or technique, describe fully the results or trend or association (using simple descriptive statistics). What do your results tell you? Describe your findings in detail by quoting the evidence from your methods of analysis.

● Describe the pattern on your graph. What does the graph show?

● Do the graphs/diagrams, etc. help to answer the question set? How?

● Make comments to link the data. For example, show how one diagram, graph or map relates to others.

● Where relevant, consider the values and attitudes of people involved.

■ Formulation of conclusions

● Using the evidence from the data you should be able to make judgements on the validity of the original hypothesis or aims of the assignment. Compare the results of the data analysis against standard models and theories.

● Link what you have discovered in your enquiry to what you have studied in the syllabus. For example, if you looked at shopping, have you talked about high-order and low-order shops, shopping hierarchies, etc.? If your investigation is on leisure, have you linked this to the amount of leisure time people have, spheres of influence of leisure centres, how accessible places are, etc.? These things should be mentioned in your introduction (the geographical theme) and need to be discussed now in relation to your findings.

● Having described your results, you need to explain and discuss them. Why have you arrived at such results? Do they confirm (accept) or refute (reject) your hypotheses?

■ Evaluation

Your project is likely to be slightly less than perfect, and you need to show the examiner that you are aware of this. It is important to realise that you will not be marked down for showing an awareness of the limitations of methods, results and conclusions. Indeed, you are more likely to be penalised if you do not include a brief section like this. There may be limits to where and when you could carry out a survey, or the number of people that you could interview. For example, there may be constraints of expense; measurements of rivers may be hampered by floods or droughts, etc.

You need to make an assessment and state whether or not these limitations have impaired your project. If they have, then your results are likely to be compromised, and so will any conclusions that you draw. For example, a survey of tourists to Oxford on a Tuesday afternoon in April found that most people were attracted to the university buildings and the colleges, and listed the poor weather as their main criticism. Had the survey taken place in another kind of tourist destination, the attractions listed would have been very different. Similarly, had the survey taken place in July or August the weather might not have been mentioned as a problem; on the other hand, congestion due to too many tourists might have been mentioned. This example shows that the methods (including the date and time of any survey) produce results that can affect our conclusions and, therefore, our evaluation. Another survey at another time would give a different set of results and conclusions.

Bibliography

At the end of the project you should make a list of any books, articles or other sources of information that you have used for your project. You do not need to include notes that you have made in class, only the extra material you used.

Case Studies

Analysing sand dunes

Sand dunes provide an interesting and manageable ecosystem for study at IGCSE. This is because significant changes can be identified over a relatively small area. In a sand dune system the most recently formed dunes are by the sea. The dunes become older with distance inland. Sand dunes form a series of ridges with intervening 'slacks' between them.

A useful starting point is to survey the morphology (size and shape) of the dunes. Figure 20a shows how measurements can be taken across a sand dune ecosystem using a tape measure and a clinometer. The transect line should be at right-

angles to the coast. The first ranging pole is carefully placed where there is a distinct break in slope from the back of the beach, marking the beginning of the sand dunes. The second ranging pole is placed at the next break of slope. The angle of slope is read from the clinometer. This process is repeated for each break of slope. With about four people working as a team, all the measurements required to draw a cross-section of the dunes (Figure 20b) can be taken in a couple of hours. If a larger group of people is available, a number of transects could be taken across the sand dunes. Transects could be compared and any differences discussed.

As the dune survey proceeds, other measurements can also be taken. At regular locations across the sand dune system the following can be measured:

- vegetation cover, with the dominant plant species noted
- maximum height of vegetation
- wind speed
- soil moisture content
- soil organic content
- soil pH.

Figure 20c provides an example of a recording sheet that could be used for such a survey. While the first few readings might take a little time, once you become familiar with what is required the process should speed up considerably.

All these measurements can be used to test the standard theories about sand dunes presented in textbooks and can be set out in a series of hypotheses to be tested:

- Vegetation density increases with distance inland.
- The number of species increases with distance inland.
- The height of vegetation increases with distance inland.
- Soil organic content increases with distance inland.
- Soil pH decreases with distance inland.
- Wind speed decreases with distance inland.

Investigating rivers

Streams and small rivers are a popular focus for geographical investigation because most schools will not be too far from a suitable example. Figure 21 (page 192) shows some of the measurements that can be taken at various locations along the course of a river. For safety reasons it is best to avoid working in a stream above the height of your knees.

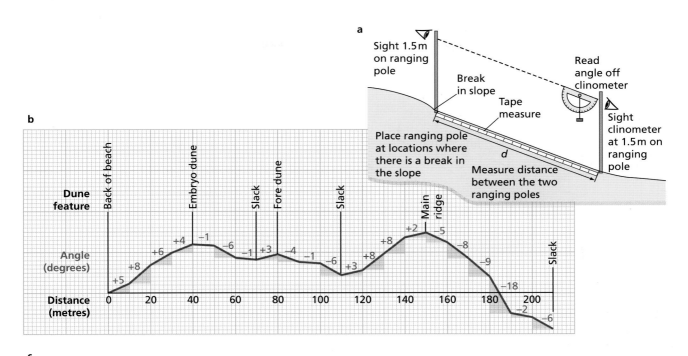

Figure 20 Conducting a survey of a sand dune ecosystem

For most river studies you will want to produce a cross-section of the river channel. The method is as follows:

1 Use a tape measure to assess the channel width. This should be done at right-angles to the course of the river. If you want to produce a cross-section of the river when discharge is at its highest you should look for evidence of the highest point the water reaches on each bank. This will give the bankfull width.

2 Channel depth should be measured at regular intervals across the river using a metre stick or ranging pole. Every 20 or 30 cm should provide an adequate sampling interval.

3 A cross-section can then be drawn using graph paper (Figure 22, page 192). As with all cross-sections, careful choice of scale is important.

Although various types of floats can be used to measure river velocity, it is best to use a flow meter. The impellor (screw device) is pointed upstream at

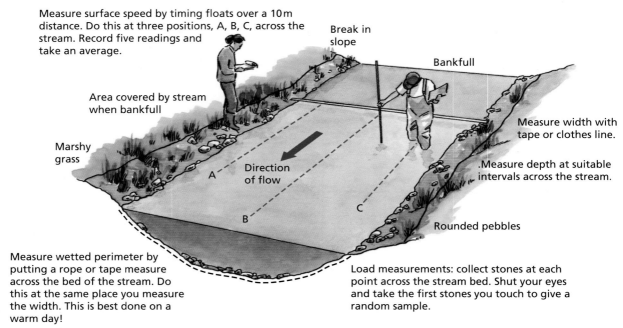

Measure surface speed by timing floats over a 10 m distance. Do this at three positions, A, B, C, across the stream. Record five readings and take an average.

Break in slope

Bankfull

Area covered by stream when bankfull

Measure width with tape or clothes line.

Marshy grass

Direction of flow

.Measure depth at suitable intervals across the stream.

A

B

C

Rounded pebbles

Measure wetted perimeter by putting a rope or tape measure across the bed of the stream. Do this at the same place you measure the width. This is best done on a warm day!

Load measurements: collect stones at each point across the stream bed. Shut your eyes and take the first stones you touch to give a random sample.

Figure 21 Taking river measurements

the same points across the river used to calculate the depth intervals. You will be able to see how velocity varies with distance from the banks and how velocity varies with depth. You could also calculate the mean flow rate for this stage of the river.

The discharge of the river can be calculated by multiplying the velocity by the cross-sectional area (Figure 22). The gradient of a river can be measured using ranging poles and a clinometer, in the same way sand dune measurements were carried out in the previous example.

Bedload measurements can also be taken to assess the impact of attrition with increasing distance downstream. Ensure that the samples of bedload are selected randomly by a ranging pole or metre stick at intervals across the river. Collect the stones that are touching the pole or stick. Measure the long axis, shape and radius of curvature of each stone.

As in the sand dune case study, a series of hypotheses based on textbook theory can be set up to be tested.

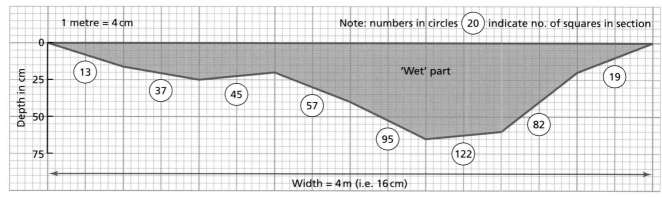

1 metre = 4 cm

Note: numbers in circles ⃝20 indicate no. of squares in section

'Wet' part

Area = 470 squares. This number has to be converted, according to the scale. The scale 4 cm = 1 m means that each 2-millimetre square on the graph paper represents 0.0025 m² in reality ($^1/_{400}$). If the scale had been 1 cm = 1 m then the scaling factor would be $^1/_{100}$ = 0.01 m². In this example 470 × 0.0025 = 1.175 m²

Figure 22 A river cross-section

Glossary of Command Words

Command words are those words in a question that tell the candidate what they have to do.

Annotate Add labels of notes or short comments, usually to a diagram, map or photograph to describe or explain.

Calculate Work out a numerical answer. In general, working should be shown, especially where two or more steps are involved.

Compare Write about what is similar and different about two things. For a comparison, two elements or themes are required. Two separate descriptions do not make a comparison.

Complete Add the remaining detail or details required.

Contrast Write about the differences between two things.

Define or **State the meaning of** or **What is meant by** Give the meaning or definition of a word or phrase.

Describe Write what something is like or where it is. **Describe** may be used for questions about resources in the question paper (describe the trend of a graph, the location of a settlement on a map, etc.). It may also be used when you need to describe something from memory (describe a meander, etc.). It is often coupled with other command words such as **Name and describe** (name the feature and say what it is like), **Describe and explain** (say what it is like and give reasons for).

Devise or **Plan** Present a particular feature such as a form or questionnaire to meet a specific requirement or requirements.

Draw Make a sketch of. Often coupled with **a labelled diagram** (draw a diagram/illustration with written notes to identify its features).

Explain or **Account for** or **Give reasons for** Write about why something occurs or happens.

Give your views or **Comment on** Say what you think about something.

How In what way? To what extent? By what means/method? May be coupled with **Show how** (prove how, demonstrate how).

Identify Pick out something from information you have been given.

Illustrate your answer Account for by using specific examples or diagrams. Often coupled with **by a labelled diagram**.

Insert or **Label** Place specific names or add details to an illustrative technique in response to a particular requirement.

Justify Say why you chose something or why you think in a certain way.

List Identify and name a number of features to meet a particular purpose.

Locate Find where something is placed or state where something is found or mark it on a map or diagram.

Measure Implies that the quantity concerned can be directly obtained from a suitable measuring instrument.

Name To state or specify or identify. To give the word or words by which a specific feature is known or to give examples that illustrate a particular feature.

Predict Use your own knowledge and understanding, probably along with information provided to state what might happen next.

Refer to or **With reference to** Write an answer that uses some of the ideas provided in a map/photograph/diagram, etc. or other additional material such as a case study.

State Set down in brief detail. To refer to an aspect of a particular feature by a short statement or by words or by a single word.

Study Look carefully at (usually one of the figures in the question paper).

Suggest Set down your ideas on or knowledge of. Often coupled with **why** (requires a statement or an explanatory statement referring to a particular feature or features).

Use or **Using the information** Base your answer on the information provided.

What Used to form a question concerned with selective ideas/details/factors.

What differences are shown between A and B Use comparative statements to describe the changes involved as A changes to B. Factual descriptions of A and B are not required.

Where At what place? To what place? From what place?

Why For what cause or reason?

With the help of information in Write an answer which uses some of the information provided as well as additional material.

Acknowledgements

The Publishers would like to thank the following for permission to reproduce copyright material:

Photo credits

Paul Guinness: **p.5**, **p.10**, **p.15**, **p.17**, **p.19** *all*, **p.21** *all*, **p.24**, **p.26**, **p.108** *all*, **p.109**, **p.111** *tr*, **p.116** *all*, **p.118**, **p.120**, **p.121** *all*, **p.122**, **p.123** *all*, **p.130**, **p.133**, **p.134**, **p.135** *all*, **p.138** *all*, **p.142**, **p.143** *tl*, **p.144**, **p.145**, **p.147**, **p.160** *br*, **p.162**.

Garrett Nagle: **p.27**, **p. 28** *all*, **p.31** *all*, **p.39**, **p.40**, **p.41**, **p.43**, **p.44** *all*, **p.47**, **p.48**, **p.52** *all*, **p.53** *all*, **p.54**, **p. 58** *all*, **p.60**, **p.62**, **p.63**, **p.66**, **p.67** *all*, **p.69** *all*, **p.70**, **p.77**, **p.78**, **p.79**, **p.81** *all*, **p.82**, **p.83** *all*, **p.84**, **p.85**, **p.92** *all*, **p.93**, **p.95** *all*, **p.96**, **p.101**, **p.102**, **p.103**, **p.104**, **p.105** *all*, **p.106**, **p.170** *all*, **p.172**, **p.177** *all*.

p.50 © David R. Frazier Photolibrary, Inc/Alamy; **p.61** © Sigurgeir Jonasson; Frank Lane Picture Agency/CORBIS; **p.110** © Arco Images GmbH/Alamy; **p.111** *tl* © V1/Alamy; **p.113** Paul Nevin/Photolibrary Group; **p.114** © Frédéric Soltan/Sygma/Corbis; **p.119** © ANTONY NJUGUNA/Reuters/Corbis; **p.128** courtesy of Volkswagen UK; **p.132** *all* Chris Guinness; **p.137** © AA World Travel Library/Alamy; **p.143** *br* JOERG BOETHLING/Still Pictures; **p.146** Invicta Kent Media/Rex Features; **p.148** © James Randklev/Corbis; **p.150** © Phillip Augustavo/Alamy; **p.154** James P. Blair/National Geographic/Getty Images; **p.155** Digital Vision/Getty Images; **p.156** © Bobby Yip/X00306/Reuters/Corbis; **p.158** © George Esiri/Reuters/CORBIS; **p.160** *tl* Caroline Schiff/The Image Bank/Getty Images; **p.162** © Anthony Kay/Flight/Alamy; **p.183** Ecoscene/Chinch Gryniewicz

Acknowledgements

p.1 *l* from P. Guinness and G. Nagle, *Advanced Geography: Concepts and Cases* (Hodder & Stoughton, 2002); *r* from Population Reference Bureau, *World Population Prospects: The 2004 Revision, 2005* (United Nations/Population Reference Bureau, 2005/2006); **p.2** *br* from Population Reference Bureau, *2007 World Population Data Sheet*; *t* from Population Reference Bureau, *2005 World Population Data Sheet* (Population Reference Bureau, 2006); **p.2** *l* from Population Reference Bureau, *World Population Prospects: 2004 Revision*, op.cit; *c* from Population Reference Bureau, *World Population Prospect: 2004 Revision*, United Nations/Population Reference Bureau, 2005/2006); **p.3** *tr & l* from *Advanced Geography: Concepts and Cases*, op.cit.; *br* Paul Guinness; **p.4** from *2007 World Population Data Sheet*, op.cit.; **p.5** *br* from *2007 World Population Data Sheet*, op.cit.; **p.6** from *Advanced Geography: Concepts and Cases*, op.cit.; **p.7** from A. Palmer and W. Yates, *Edexcel (A) Advanced Geography* (Philip Allan, 2005); **p.8** *l* from *Advanced Geography: Concepts and Cases*, op.cit.; *r* from Population Reference Bureau, *Population Bulletin*, Volume 60, No.4 (Population Reference Bureau, 2005); **p.9** *all* from *Population Bulletin*, Volume 60, No.4 (Population Reference Bureau, 2005); **p.10** *l* from Paul Guinness and G. Nagle, *Geocases* (Hodder Murray, 2006), reprinted by permission of the publisher; **p.11** from *International Data Base* (US Census Bureau, 2007); **p.12** from M. Harcourt and S. Warren, *Tomorrow's Geography* (Hodder Murray), reprinted by permission of the publisher; **p.13** *tr* from *www.12.statcan.ca/english/census06/ analysis/agesex/charts/chart21.htm*, reprinted by permission of Statistics Canada; *br* Paul Guinness; **p.14** from *2007 World Population Data Sheet*, op.cit.; p.15 *l* from *2007 World Population Data Sheet*, op. cit.; **p.16** from M. Carr, *New Patterns: Process and Change in Human Geography* (Nelson Thornes, 1999); **p.17** *tl* from *Advanced Geography: Concepts and Cases*, op.cit.; *b* G. Nagle and K. Spencer, *Advanced Geography: Revision Handbook* (Oxford University Press, 1996), reprinted by permission of the publisher; **p.18** from Tom Stevenson,

Daily Telegraph (5 September 2007), reprinted by permission of Telegraph Media Group; **p.19** *b* from P Guinness and G. Nagle, *AS Geography: Concepts and Cases* (Hodder Murray, 2000), reprinted by permission of the publisher; p.20 *l* from *2007 World Population Data Sheet*, op. cit; *r* from *Geofile Online*, No. 429 (Nelson Thornes, September 2002); **p.21** *tl* from *Advanced Geography: Concepts and Cases*, op.cit.; **p.23** *all* from *Advanced Geography: Revision Handbook*, op.cit.; **p.24** *t* from *Geocases I* (Hodder Murray), reprinted by permission of the publisher; *b* from *2007 World Population Data Sheet*, op.cit; **p.25** *all* from *Newsweek*, 19/01/04; **p.26** *tr* from M. Parnwell, *Population Movements and the Third World* (Routledge, 1993), reprinted by permission of Taylor and Francis Books UK; **p.28** *bl* from G. Nagle, *Advanced Geography* (Oxford University Press, 2000), reprinted by permission of the publisher; **p.29** *t* G. Nagle; **p.30** *t* from *Advanced Geography: Concepts and Cases*, op.cit; *b* from G. Nagle, *Geography Homework Pack for Key Stage 3* (Heinemann, 2000); **p.31** *b* from G. Nagle, *Geography Through Diagrams* (Oxford University Press, 1998), reprinted by permission of the publisher; **p.33** from Government of Jamaica Survey Department; **p.34** from Government of Jamaica Survey Department, op.cit.; **p.35** from G. Nagle, *Thinking Geography* (Hodder Murray, 2000), reprinted by permission of the publisher; **p.36** from *Thinking Geography*, op.cit.; **p.37** from S. Warn, *Managing Change in Human Environments* (Philip Allan Updates, 2001); **pp.39-41** from New York City, Department of City Planning, *www.nyc.gov/html/dcp/html/ landusefacts/landusefactsmaps.html*; **p.42** *all* from Young-Han Park et.al., *Atlas of Seoul* (Sung Ji Mun Hva Co. Ltd, 2000); **p.44** *tr* G. Nagle; **p.45** from *The Challenge of Slums: UN-HABITAT'S Global Report on Human Settlements* (2003); **p.46** from *GeoFactsheets*, 121; **p.47** *r* G. Nagle; **p.48** *tl* from G. Nagle, *AS and A2 Geography for Edexcel B* (Oxford University Press, 2003); *b* from *Managing Change in Human Environments*, op.cit; **p.49** *t* from *Managing Change in Human Environments*, op.cit; *bl & br* from G. Nagle, *AS and A2 Geography for Edexcel B* (Oxford University Press, 2003), reprinted by permission of the publisher; **p.51** from *GeoFactsheets*, 121, op.cit; **p.52** *t* from *Managing Change in Human Environments*, op.cit; **p.53** *b* G. Nagle; p. 55 from *AS Geography: Concepts and Cases*, op. cit.; p. 56 *t* from *AS Geography: Concepts and* Cases, op. cit.; p. 57 from *AS Geography: Concepts and Cases*, op.cit; **p.59** *all* from *Thinking Geography*, op.cit.; **p.60** *b* from Directorate of Overseas Surveys; **p.63** *b* from G. Nagle, *Hazards* (Nelson Thornes, 1998); **p. 64** from *Philip's Modern School Atlas*, 95th edition (George Philip Maps Ltd.); **p.65** G. Nagle; **p.68** *t* from *Advanced Geography*, op.cit.; *b* from *Geography Through Diagrams*, op.cit.; **p.71** *t* from *Advanced Geography*, op.cit.; *b* from *AS and A2 Geography for Edexcel B*, op.cit.; **p.72** *bl* from *Advanced Geography*, op.cit.; *br* from G. Nagle, *Rivers and Water Management* (Hodder Arnold, 2003), reprinted by permission of the publisher; **p.73** from Canada Map Office, Ottawa; **p.74** *l* from *Geography Through Diagrams*, op.cit.; *r* from *AS Geography: Concepts and Cases*, op.cit.; **p.75** from *AS and A2 Geography for Edexcel B*, op.cit.; **p.76** *l* from *AS and A2 Geography for Edexcel B*, op.cit.; *r* from *AS Geography: Concepts and Cases*, op.cit; **p.77** *tl* from *Geography Through Diagrams*, op.cit.; *bl* from Department: Land Affairs, Republic of South Africa, reprinted by permission of Chief Directorate: Surveys and Mapping; **p.78** *t & c* from *Geography Through Diagrams*, op.cit.; **p.79** *tr* from *www.itmb.com*, reprinted by permission of ITMB Publishing; p. 80 from *AS Geography: Concepts and Cases*, op. cit.; **p.82** *b* from Sue Warn and Carol Roberts, *Coral Reefs: Ecosystem in Crisis?* (Field Studies Council, 2001), reprinted by permission of the publisher; **p.84** *b* from G. Nagle, *Weatherfile GCSE* (Nelson Thornes, 2000); **p.87** *b* from *Weatherfile GCSE*, op.cit.; **p.89** *all* from *Philip's Modern School Atlas*, 95th edition (George Philip Maps Ltd.); **p.90** from *Thinking Geography*, op.cit.; p. 90 from *AS Geography: Concepts and* Cases, op.cit.; **p.92** *b* from *Geography Through Diagrams*, op.cit.; **p.97** *t* from P. Guinness and G. Nagle, *Geography for CSEC* (Nelson Thornes, 2008); *b* from *Philip's Certificate Atlas for the Caribbean*, 5th Edition (George Philip Maps, 2004); **p.98** from P. Guinness and G. Nagle, *Geocases 4* (Hodder Murray, 2005), courtesy: NOAA National

Hurricane Center; **p.99** from *AS Geography: Concepts and Cases,* op.cit.; **p.100** from *AS Geography: Concepts and Cases,* op.cit; **p.107** from P. Guinness, *North America in Focus* (Hodder and Stoughton, 1990), reprinted by permission of the publisher; **p.108** *t* from David Waugh, *Geography: An Integrated Approach,* 1st Edition (Nelson Thornes, 1990); **p.109** *b* from *Geography: An Intergrated Approach,* op.cit.; **p.111** *b* from *North America in Focus,* op.cit.; **p.112** *http://wool.com.au/Pastures/Introduction_to_Australian_Pastures/ page_2122.aspx* , Land and Water Resources Audit; **p.114** *t & bl* from David Waugh, *The New Wider World,* 2nd Edition (Nelson Thornes, 2003); **p.115** from Julia Waterlow, *The Amazon* (World's Rivers) (Hodder Wayland, 1992); **p.117** from *Geofactsheet,* No. 185 (Curriculum Press, September 2005), reprinted by permission of the publisher; **p.118** *t* from *Geofactsheet,* No. 185 op.cit.; **p.119** *t* from *Geofactsheet,* No. 185, op.cit.; **p.122** *tl* from *Advanced Geography: Concepts and Cases,* op.cit.; **p.123** *b* from *The New Wider World,* 2nd Edition, op.cit.; **p.125** *b* from *http://city.ottawa.on.ca,* reprinted by permission of Communications and Customer Service Branch, City of Ottawa; **p.126** *t* from *www.ocri.ca/economicstatistics/htsurvey.asp*; *b* from CE Info Systems, © 2008; **p.127** from *Geofile,* No.505 (Nelson Thornes, September 2005); **p.128** *r* P. Guinness; **p.130** *b* from World Tourism Organisation (UNWTO); **p.131** *b* from World Tourism Organisation (UNWTO), op.cit.; **p.133** *l* from Global Insight: Tourism Satellite Accounting, reprinted by permission of Global Insight; **p.134** *l* P. Guinness; **p.136** from Jane Dove et.al., *OCR AS Geography* (Heinemann Educational, 2008); **p.139** from *Geofactsheet,* No. 201, reprinted by permission of Curriculum Press; p. 140 *b* from *OCR AS* Geography, op.cit.; **p.141** *bl* from DNE National Directorate of Energy, Mozambique; **p.142** *l & br* from *BP Statistical Review of World Energy* (BP, June 2007); **p.144** *r* from *Financial Times* (20 June 2007), reprinted by permission of the publisher; *tl* from *Global Wind 2007 Report* (Global Wind Energy Council), reprinted by permission of the publisher; **p.145** *tl* from *North America in Focus,* op.cit; *r* from Paul Guinness, *Resources and Power* (Macdonald Publishers, 1984); **p.146** *l* from *www.hknuclear.com* (Hong Kong Nuclear Investment Company), reprinted by permission; **p.147** *l* from *EarthTrends* (World Resources Institute, 2007), reprinted by permission of WRI; **p.148** *b* from *Advanced Geography: Concepts and Cases,* op.cit.; **p.149** from *Advanced Geography: Concepts and Cases,* op.cit.; **p.152** from *North America in*

Focus, op.cit.; **p.153** *t* from *Advanced Geography: Concepts and Cases,* op.cit.; *b* from J. Hill, W.Woodland, R.Hill, *Geography Review* (Geography Review, May, 2007); **p.154** *l* from *www.virtualsources.com/countries/Latin%20/Countries/ar-map.gif*; **p.156** *l* from *Financial Times* (4 February 2003), reprinted by permission of the publisher; **p.157** from *Daily Telegraph* (29 April 2008), reprinted by permission of Telegraph Media Group; **p.158** *l* from *National Geographic* (National Geographic Society, February 2007); p. 159 from *OCR AS Geography,* op.cit.; **p.160** *tr* from *http://www.great-barrier-reef.biz/Images/100050.gif* **p.161** from DEFRA, © Crown copyright; **p.162** *l* from *Evening Standard* (21 February 2008), reprinted by permission of Solo Syndication; **p.163** from Overseas Surveys Directorate, Ordnance Survey; **p.166** from Ordnance Survey on behalf of HMSO, © Crown copyright (Client Licence Number: 100036470); **p.167** from G. Nagle and K.Spencer, *Geographical Enquiries* (Stanley Thornes, 1997); **p.168** from *www.itmb.com*; **p.169** *all* from *www.itmb.com*; **p.171** from *www.theAA.com/travel,* © KOMPASS-Karten Gmbh., reprinted by permission of KOMPASS; **p.173** from Canada Map Office, Ottawa; **p.174** from *Korea Statistical Yearbook* (2000); **p. 175** *tl and bl* from *Korea Statistical Yearbook* (2006); *r* P. Guinness and G. Nagle; **p.176** *t* from *Advanced Geography.,* op.cit; *b* from *Academy of Korean Studies*; **p.182** from B. Lenon and P. Cleves, *Fieldwork Techniques and Projects in Geography* (Collins Educational, 2001); **p.183** *t* from *Wideworld* (Philip Allan Updates, November 2004); **p.184** from *Wideworld,* November 2004, op.cit.; **p.185** *b* from *Wideworld,* November 2004, op.cit.; *t & c* P. Guinness; **p.186** *l* P. Guinness; *tr* P. Guinness; *br* from *Fieldwork Techniques and Projects in Geography,* op.cit.; **p.187** *l* from *Wideworld* (Philip Allan Updates, November 2001); **p.188** *l* from *Fieldwork Techniques and Projects in Geography,* op.cit.; *tr* P. Guinness; *br* from *Wideworld* (Philip Allan Updates, February 2004); **p.189** from *Wideworld* (Philip Allan Updates, November 2002); **p.190** from *Wideworld* (Philip Allan Updates, September 2002); **p.191** from *Wideworld,* September 2002, op. cit; **p.192** all from D. Holmes and S. Warn, *Fieldwork Investigations: a self-study guide* (Hodder & Stoughton), reprinted by permission of the publisher.

Every effort has been made to trace all copyright holders, but if any have been inadvertently overlooked the Publishers will be pleased to make the necessary arrangements at the first opportunity.